SpringerBriefs in Computer Science

SpringerBriefs present concise summaries of cutting-edge research and practical applications across a wide spectrum of fields. Featuring compact volumes of 50 to 125 pages, the series covers a range of content from professional to academic.

Typical topics might include:

- A timely report of state-of-the art analytical techniques
- A bridge between new research results, as published in journal articles, and a contextual literature review
- A snapshot of a hot or emerging topic
- An in-depth case study or clinical example
- A presentation of core concepts that students must understand in order to make independent contributions

Briefs allow authors to present their ideas and readers to absorb them with minimal time investment. Briefs will be published as part of Springer's eBook collection, with millions of users worldwide. In addition, Briefs will be available for individual print and electronic purchase. Briefs are characterized by fast, global electronic dissemination, standard publishing contracts, easy-to-use manuscript preparation and formatting guidelines, and expedited production schedules. We aim for publication 8–12 weeks after acceptance. Both solicited and unsolicited manuscripts are considered for publication in this series.

More information about this series at http://www.springer.com/series/10028

Brian Kokensparger

Guide to Programming for the Digital Humanities

Lessons for Introductory Python

Brian Kokensparger
Department of Journalism,
Media and Computing
Creighton University
Omaha, NE, USA

ISSN 2191-5768 ISSN 2191-5776 (electronic)
SpringerBriefs in Computer Science
ISBN 978-3-319-99114-6 ISBN 978-3-319-99115-3 (eBook)
https://doi.org/10.1007/978-3-319-99115-3

Library of Congress Control Number: 2018950953

This Springer imprint is published by the registered company Springer Nature Switzerland AG
The registered company address is: Gewerbestrasse 11, 6330 Cham, Switzerland

Contents

Chapter 1
Introduction

Abstract An introduction to the topic of using Digital Humanities assignments for an introductory programming course in Python, which lays out the context of the book. It begins with a brief anecdote about what led the author to initially develop these assignments, provides a brief overview of the field of Digital Humanities, and introduces the reader to how the book is arranged with a hint of what is to come.

Keywords Digital-humanities assignments · Introductory programming
Python course · Digital-humanities origins

1.1 A Call to the Digital Humanities

Imagine that you are a computer science instructor, moving happily along in your career. You teach the same old programming courses: introductory programming, introduction to computing, and computer organization. Your syllabi are set, and you barely need to review your notes before shuffling into class each day. Then one day, an associate dean calls: the new Digital Humanities (DH) initiative needs a Programming for the Humanities course. And wouldn't you like to teach it?

This happened to me. What should I teach? How should I teach it? How is programming for the humanities different from any other programming?

The solution, for me, was to teach all the concepts and structures typical for a regular computer science course (loops, branching structures, variables for assignment, etc.), but to incorporate programming assignments from the digital humanities domain (textual analysis, data visualization, stylometry, social network analysis, etc.). This book provides the full instructions, sample code, and resources for current introductory programming instructors to adopt these DH assignments into their own courses using the Python language.

All the assignments in this book have been used over at least 4 semesters in my classroom, in classes that average from 26 to 30 students. The assignments were originally designed for the course, and have been tweaked and assessed over time to optimize student learning. This is a tricky process, because the range of students

B. Kokensparger, *Guide to Programming for the Digital Humanities*, SpringerBriefs in Computer Science, https://doi.org/10.1007/978-3-319-99115-3_1

who arrive in an introductory programming course varies from those with no coding experience at all (common especially among the humanities students), to those who already have impressive programming skills. Each of these assignments are presented specifically to be used with novice programming students, but each also includes a discussion of the flexibility allowed for students who have higher-level skills. The assignments have been successfully implemented for this entire range of student abilities.

The assignments include: Computing Change over Time (calculating burials in a historic cemetery), Visualization of Change over Time (visualizing the burials in the historic cemetery), Textual Analysis (finding word frequencies and "stop words" in public domain texts), XML Transformation (programmatically transforming a simplified version of an XML file into an HTML file, styled with CSS), Stylometry (comparing measured features of graphic images), and Social Network Analysis (analyzing extended relationships in historic circles). In each of these assignments, the text presents the general domain (DH significance) of the assignment, the programming concepts and skills that the assignment will enhance and reinforce, a few variations that can be applied based on the programming skill of the students, and some hints and tips for successfully introducing, managing, and grading the submitted work.

1.2 Brief Overview of Digital Humanities

I have developed this overview from some of my previously published papers. Digital humanities (DH) began along with the mainstream implementation of computers in the late 1940s with conceptual projects, followed by efforts to digitize and archive large volumes of texts in the 1970s [4]. As computing became more accessible through the development of the personal computer (in the mid-1980s), so did the availability and, in succession, digitization of humanities data [3]. The rise of the World Wide Web in the 1990s "accelerated the transition in digital scholarship from processing to networking" [4]. One source had even asserted that "the Digital Humanities may well function as a core curriculum for the 21st Century" [4].

As DH "remains at its core a profoundly collaborative enterprise" [4, 6, 8], and by its definition it consists of the application of computer technology to humanities research [5], it is no surprise that university deans have called upon computer science departments to provide programming support and instruction for DH initiatives [10]. These initiatives often are successful at attracting female students to computer science [1], as well as students from other demographic categories who are difficult to attract and engage [9], and is a good way to engage students in institutions that already have thriving cultures in the humanities [2].

This was the case at my institution, Creighton University. We had begun a DH initiative in 2016, and needed a programming for the humanities course to fill out a DH minor. This approved minor now includes 18 semester hours of credit in DH program-approved courses—one of them being the required programming for the humanities course, listed in the curriculum requirements as "CSC 221: Introductory Programming".[1]

As one of the few faculty members in my department who had a humanities background (a B.A. in English with a good solid background in philosophy and theology), I took on the task of developing an introductory programming course into a dual-role course that also served as programming for the humanities. It was the dual nature of this course that prompted me to teach all of the normal introductory programming topics, but to simultaneously employ DH assignments to give DH students, in particular, practical experience in programming for the humanities.

This approach allows the students who are only pursuing computer science (CS) majors to learn all of the basic concepts normally taught in introductory programming courses (and thus prepare them fully for their next programming course) [7], but also allows DH students to gain familiarity with how programming creates the tools they will be using as future professionals in their field.

Lest you think that the CS students are missing out on the normal business-related or computationally-related programming experiences, the reality of our curriculum is that they will get the opportunity to gain these experiences in other courses that they take later in their undergraduate studies in the program, whereas this may be the only programming course that the DH students will take.

1.3 How This Book Is Arranged

This book is arranged for you to enter into the assignments by first discussing the curriculum and topics covered in typical introductory programming courses (Chap. 2), followed by a discussion of the special considerations for digital humanities programming (Chap. 3) and then an introduction to the digital humanities assignments as a whole (Chap. 4).

The chapters that follow present the materials for each of the assignments: Change over Time (Chap. 5), Visualizing Change over Time (Chap. 6), Textual Analysis (Chap. 7), Code Transformation (Chap. 8), Art Stylometry (Chap. 9), and Social Network Analysis (Chap. 10).

A conclusion and thoughts for additional assignments are provided in the final chapter (Chap. 11).

1.4 Only the Beginning

My story is still writing itself, but since student satisfaction with the course (including its assignments) is high—including evaluations from among the DH students—I have great confidence that these assignments will work in your classroom as well. If you

[1] https://www.creighton.edu/program/digital-humanities-minor.

are a computer science instructor, I have written this book to share these assignments with you, whether you are either required or self-motivated to adopt DH assignments into your introductory programming course.

Or, if you have picked up this book as a basis for self-instruction, I am also confident that the assignments presented here will give you a good start in your own journey to find the data in the Human, and the Human in the data. Don't quite understand what I just wrote? You will be the end of this book.

So we begin, in the next chapter, with a look into the topics for a typical introductory programming (CS1) course.

References

1. Alvarado, C., Dodds, Z.: Women in CS: an evaluation of three promising practices. Paper presented at the 41st ACM technical symposium on computer science education, Milwaukee, Wisconsin, 10–13 March 2010
2. Beck, R.E.: Digital humanities: reaching out to the other culture. Paper presented at the 43rd ACM technical symposium on computer science education, Raleigh, North Carolina, 29 February–3 March 2012
3. Berry, D.M.: The computational turn: thinking about the Digital Humanities. Culture Machine **12**, 2 (2011)
4. Burdick, A., Drucker, J., Lunenfeld, P., Presner, T., Schnapp, J.: Digital Humanities. The MIT Press, Cambridge (2012)
5. Cassel, L.B., MacKellar, B., Peckham, J., Spradling, C., Reichgelt, H., Westbrook, S., Wolz, U.: Interdisciplinary computing in many forms. Paper presented at the 45th ACM technical symposium on computer science education, Atlanta, Georgia, 5–8 March 2014
6. LeBlanc, M.D., Drout, M.D.: DNA and 普通話 (Mandarin): bringing introductory programming to the Life Sciences and Digital Humanities. Procedia Comput. Sci. **51**, 1937–1946 (2015)
7. Powers, K., Gross, P., Cooper, S., McNally, M., Goldman, K.J., Proulx, V., Carlisle, M.: Tools for teaching introductory programming: what works?. In: ACM SIGCSE Bulletin, vol. 38, No. 1, pp. 560–561. ACM (2006)
8. Schreibman, S., Siemens, R., Unsworth, J. (eds.).: A companion to Digital Humanities. Wiley, Hoboken (2008)
9. Siiman, L.A., Pedaste, M., Tõnisson, E., Sell, R., Jaakkola, T., Alimisis, D.: A review of interventions to recruit and retain ICT students. Int. J. Mod. Educ. Comput. Sci. (IJMECS) **6**(3), 45 (2014)
10. Vee, A.: Understanding computer programming as a literacy. Literacy Compos. Stud. **1**(2), 42–64 (2013)

Chapter 2
Introductory Programming—Common Topics

Abstract This chapter discusses topics normally taught within the typical introductory programming course (also known as a CS1 course among computer science educators) for undergraduate college students, including topics suggested by the ACM Computer Science Curricula 2013 report. It also provides learning "targets," or single-point assessible skills and knowledge, for all the skills related to the normal topics taught in an introductory programming course, such as Introduction to Computer Basics and Programming, Accessing the IDE and Basic Python Syntax, Variables, Numbers, and Expressions, Instantiating Objects and Introduction to Graphics in Python, Working with Strings and other Sequence Structures, Using Pre-Defined Functions and Creating User-Defined Functions, Creating and Testing Branching Structures, Creating and Managing Looping Structures, Working with Lists and Arrays, Program Algorithms and the Design Process, and Brief Introduction to Object-Oriented Programming in Python, as well as some general suggestions about approaches to teaching these topics.

Keywords Introductory programming · Curricula 2013 · Learning targets
Python topics · Programming topics

2.1 The Introductory Programming Curriculum

Though the scope of this book is not to detail the common topics presented in an introductory programming course, it is important for those who may be coming at this book from a novice instructor's perspective. If you have already taught introductory programming as a course, please feel free to skip this chapter and move on to the next, where a brief introduction to the special considerations for teaching Digital Humanities awaits.

Electronic supplementary material The online version of this chapter (https://doi.org/10.1007/978-3-319-99115-3_2) contains supplementary material, which is available to authorized users.

B. Kokensparger, *Guide to Programming for the Digital Humanities*, SpringerBriefs in Computer Science, https://doi.org/10.1007/978-3-319-99115-3_2

For the rest of us, however, I can only share the course topics I teach as a way of discussing those common to any introductory programming course. There is no doubt in my mind that if I gather a group of 30 computer science instructors teaching an introductory programming course, they will produce 30 unique topic lists among their courses. However, I am equally sure that there will easily be an intersection among all of those topics among most if not all 30 of the instructors. It is those most common topics that I will present here in this chapter.

You should be aware, however, that my order of topics might be different than yours. I also take the strategy (in agreement with my chosen textbook for the introductory programming in Python course) that some introduction to later concepts at an earlier part of the semester is important to student learning (called *iterative learning* by some). In this regard, you might see this approach as running a familiarity/application/mastery spectrum, but staying between familiarity and application for the most part, since this is an introductory programming course. With that being said, here is the catalog course description of the introductory course I teach at Creighton University: "A first course in computer programming and problem solving, with an emphasis on multimedia applications. Specific topics include algorithm development, basic control structures, simple data types and data structures, and image/sound processing" [3].

As instructors, we are free to tweak the course a bit, and my tweak is away from multimedia applications (though I do a good bit of image processing, as you will see) and towards digital humanities applications. The only part of the course description that I leave out in my current offering of the course is sound processing, though there is an area in digital humanities that does focus on that topic [2], so there is a chance that I may work that in at a later date, to be fully compliant with the description.

2.2 The Topics, as Suggested by the Curricula 2013 Report

Luckily, the topics for the syllabus from which I teach were included in the CS Computer Science Curricula 2013 report [1], so some additional detail for the topics are included in Fig. 2.1. The Knowledge Area (KA) identifiers (on page 450 of the document) [1] are Software Development Fundamentals (SDF), Programming Languages (PL), Algorithms and Complexity (AL), and Social Issues and Professional Practice (SP).

How I typically integrate these general topics in an introductory programming course, using a 15-week semester offering, is included below, expressed as learning targets for each major topic. A learning target is a student learning sub-objective, for which it is easy to build a simple assessment instrument, like a single test question.

Body of Knowledge coverage

KA	Knowledge Unit	Topics Covered	Hours
SDF	Algorithms and Design	concept & properties of algorithms; role of algorithms in problem solving; problem solving strategies (iteration, divide & conquer); implementation of algorithms; design concepts & principles (abstraction, decomposition)	8
SDF	Fundamental Programming Concepts	syntax & semantics; variables & primitives, expressions & assignments; simple I/O; conditionals & iteration; functions & parameters	8
SDF	Fundamental Data Structures	arrays; records; strings; strategies for choosing the appropriate data structure	8
SDF	Development Methods	program correctness (specification, defensive programming, testing fundamentals, pre/postconditions); modern environments; debugging strategies; documentation & program style	6
PL	Object-Oriented Programming	object-oriented design; classes & objects; fields & methods	3
PL	Basic Type Systems	primitive types; type safety & errors	1
PL	Language Translation	interpretation; translation pipeline	1
AL	Fundamental Data Structures and Algorithms	simple numerical algorithms; sequential search; simple string processing	4
SP	History	history of computer hardware; pioneers of computing; history of Internet	1

Fig. 2.1 Computer Science Curricula 2013 outline of topical coverage in Creighton's CSC 221 course, found on page 451 [1]

2.3 Topics of a Typical Introductory Programming Course in Python

2.3.1 Introduction to Computer Basics and Programming

- articulate the difference between hardware and software;
- define computer science; define the term *analysis*;
- situate the Python language in terms of high-level/low-level, and compiled/interpreted; open and use the IDE (integrated development environment);
- create a new Python module (file) that can be saved, opened, and run at will;
- print a blank line;
- print the result of an expression of your choice;
- define and run a simple hello() function (from given code);

- define and run a simple greet(person) function (from given code);
- make a mistake and debug it;
- create a simple Python module that uses these concepts and skills to solve a given problem.

2.3.2 Accessing the IDE and Basic Python Syntax

- list and describe the steps of the software development process;
- articulate the rules of choosing identifier names (for variables and functions);
- give examples of legal and illegal identifiers;
- create a variable and assign a literal value to it;
- create a variable and assign the result of an expression to it;
- use an input function call to request keyboard entry from the user and assign it to a variable;
- use simultaneous assignment to assign literal values to two variables;
- create a simple for-loop (from given code) using a call to the range function to repeat a statement a given number of times.

Assignment 1—Change Over Time—is normally assigned after this topic has been covered.

2.3.3 Variables, Numbers, and Expressions

- identify the primary built-in data types;
- list all of the built-in numeric operators and the operation each performs;
- list all of the type conversion functions;
- given a math library function (like cos(x)), tell what the function receives as input and provides as a return value;
- define a variable iteratively adjusted in a loop and provide an example of how and when it may be used;
- use the type function to determine the data type of a variable or literal value;
- appropriately use the /, //, and % operators;
- convert a floating point value into an integer;
- convert an integer into a floating point number;
- use the round() function appropriately;
- import and use the math function library;
- use the math library functions ceil() and floor() appropriately.

2.3.4 Instantiating Objects and Introduction to Graphics in Python

- list and describe the steps of the software development process;
- discuss how the dot (.) notation is used to identify objects, attributes, and methods;
- give the reasons why it is better to use a clone() function instead of a simple assignment statement to make a copy of an existing object;
- import Zelle's graphics.py library[1] of classes (or another graphics library of your choice);
- find the online documentation for the graphics.py library and instantiate an object from a given class;
- create a GraphWin object and use it to display a geometric object of your choice;
- create Point(), Line(), Circle(), Rectangle(), and Text() objects and draw them on a graphic window;
- use the clone() function to correctly copy an object;
- get a mouse click from a graphics window;
- get text input from a graphics window.

Assignment 2—Visualizing Change Over Time—is normally assigned after this topic has been covered.

2.3.5 Working with Strings and Other Sequence Structures

- identify the primary difference between a string and other built-in data types;
- define the term indexing and how it relates to string data;
- define the term slicing and how it relates to string data;
- list the string operators and the tasks that they perform on string data;
- given a specific string method, describe what the method does and the method's input and returned value;
- discuss the basic concept of file input and output, including a general explanation of what happens in computer memory during read and write operations;
- create a variable and assign a literal string value to it;
- given an index value, write an expression that returns the character at that index;
- given a range, write an expression that returns the character(s) within that range;
- provided a given task and a string, find the appropriate string method for the task and write an expression that returns the appropriate value;

[1] Dr. Zelle's excellent graphics library can be found at http://mcsp.wartburg.edu/zelle/python/, which provides a link to his graphics.py module and documentation. If you do not use this graphics library, there are other libraries you can use, or develop your own using the TKInter library (information available at https://wiki.python.org/moin/TkInter).

- create a text file (outside of Python), open the file for reading, and read in the contents of the file;
- open a file for writing, and, using the *print*() method, write out given contents to the file.

2.3.6 Using Pre-Defined Functions and Creating User-Defined Functions

- provide a number of reasons why functions are helpful to the Python programmer;
- articulate the basic parts of a function definition and function call, including: function heading, function name, parameter list, body (statements) of the function, and function call;
- describe the difference between pass-by-value and pass-by-reference for parameters, and which one Python uses by default;
- define and call a function with no parameters and no returned value;
- define and call a function with one or more parameters and no returned value;
- define and call a function with no parameters and a returned value;
- define and call a function with one or more parameters and a returned value;
- define and call a function with a list as a parameter that changes in the function body.

Assignment 3—Textual Analysis—is normally assigned after this topic has been covered.

2.3.7 Creating and Testing Branching Structures

- list and describe the use of the 6 relational operators;
- articulate the purpose of decision structures and when and how to use them;
- define semantic errors and how decision structures particularly are subject to them;
- define runtime errors and how exception handling protects a program from them;
- form a given boolean expression that produces an output of True or False in the IDE shell or console;
- find the data type of the values True and False;
- find the data type of the values true and false;
- write a proper 2-way (if) decision structure; write a proper 3-way (if-else) decision structure;
- write a proper multi-way (if-elif-else) decision structure;
- from an example, write a simple exception handling routine to capture a specific *ErrorType*.

2.3.8 Creating and Managing Looping Structures

- identify the difference between definite and indefinite loops, and provide loop names that represent each;
- discuss in plain language the construction of each of the common loop patterns provided in class;
- define a nested loop and when and how they are used;
- identify the boolean operators and when and how they are used;
- write a counted/definite loop (for-loop);
- write an interactive loop;
- write a sentinel loop;
- write a file loop; write a nested loop;
- given a logical construction, create a boolean expression to test the construction.

Assignment 4—File Transformation—is normally assigned after this topic has been covered.

2.3.9 Working with Lists and Arrays

- articulate the similarities and differences between strings and lists;
- name the various list operations, and what they do;
- given a list method, tell what the method does, its parameter value (if any), and its return value (if any);
- write code that concatenates two lists and assigns the result to a variable;
- given a specific index value to a list, write code that assigns the value at that index to a variable;
- write code that assigns the length (number of elements) of a list to a variable;
- write code that traverses the entire list, printing each of its values;
- write code that traverses the entire list using the *range* function, printing each of its values;
- write code that returns a boolean value if a specific value is in a list;
- given a number of list methods, use that method to perform the given task on a list.

Assignment 5—Art Stylometry —is normally assigned after this topic has been covered.

2.3.10 *Program Algorithms and the Design Process*

- explain why "random" numbers produced by a computer can never be truly random;
- describe the steps to perform top-down design, including the strengths of that approach;
- describe the steps to perform spiral design, including the strengths of that approach;
- create a Python module by using one of the design approaches that uses these concepts (as well as concepts and skills learned in previous chapters) to solve a given problem.

2.3.11 *Brief Introduction to Object-Oriented Programming in Python*

- review the items previously learned in the semester regarding creating and using objects instantiated from classes;
- give several reasons why object-oriented programming offers the programmer an attractive alternative to functional programming;
- given a real-world object, design, define, and implement a simple class to model the object in Python code;
- having created a class, instantiate an object of the class and call its various methods to test it.

Assignment 6—Social Network Analysis—is normally assigned after this topic has been covered.

2.4 Basic Approaches to Teaching the Topics

As some of you may have noted, this approach closely follows the structure set up by Dr. John Zelle in his textbook "Python Programming: An Introduction to Computer Science" [4]. This topical progression, and the assignments presented in Chaps. 5–10 of this book, work well together in keeping the challenges accessible, with the goal of distributing the assignments evenly over an entire semester.

However, if you are using a different textbook, it is quite possible to rearrange the assignments—or the chapter topics as presented in the textbook—to work out a good solid semester of experiences for any student. The "Skills Utilized in this Assignment" section, included in each assignment chapter of this book, provides a list of the skills practiced with each programming assignment, which may be helpful to you when working with other textbooks while you deploy these assignments.

And, of course, you are perfectly at liberty to alter these assignments to suit your own approach to the topics and material.

The DH focus of this book, and the assignments in my course, are designed to fully support—and not disrupt—learning introductory programming concepts and techniques. In fact, they give introductory programming students something "real" to do, which in my opinion is rare at the introductory level, where trivial programming assignments are often offered.

The only caveat I present is that there is a chance that students may become so enthralled with the DH aspect of the course that they lose out on learning the fundamental concepts of programming. This could happen if lots of starter code and a generous number of hints are provided, to a point where the instructor is—for all intents and purposes—giving the solution code away to the students. Even a class full of DH students deserves to have a solid foundation of programming skills in an introductory programming course. To deny this to the class is a tragedy from both the CS and DH perspectives.

Computer Science needs fundamentally good programmers among its ranks; indeed, students who are not fundamentally sound have very little chance of successfully completing their CS curricula in higher education programs, and being successful professionals in the field.

Furthermore, DH also needs fundamentally good programmers among the ranks of its professionals. Your DH students will get plenty of exposure to DH methodology in other coursework and project work, but this may be the only chance they will have at learning the fundamental concepts of programming a computer. It is important to keep this priority in mind as you build and deliver an introductory programming course that is, secondarily, focused on programming for the humanities.

In the next chapter, we will briefly survey this secondary course goal—but primary goal of this book.

References

1. ACM IEEE-CS Joint Task Force: Computer science curricula 2013: final report. https://www.acm.org/binaries/content/assets/education/cs2013_web_final.pdf (2013). Accessed 19 June 2018
2. Barber, J.F.: Sound and digital humanities: reflecting on a DHSI course. Digital Humanit. Q. **10**(1) (2016)
3. Computer Science and Informatics. Creighton University, Omaha, Nebraska. http://catalog.creighton.edu/undergraduate/arts-sciences/journalism/computer-science-informatics-computing-science-track-bs/ (2018). Accessed 25 June 2018
4. Zelle, J.: Python Programming: An Introduction to Computer Science, 3rd edn. Franklin, Beadle, Portland, Oregon (2016)

Chapter 3
Digital Humanities—Special Considerations for the Programmer

Abstract The field of Digital Humanities, in dealing with data produced by humans about humans (i.e., the Humanities), requires some special considerations for the programmer. This chapter provides a brief description of some popular analytical approaches to the digital humanities, including Digital Archiving and Editing, General Quantitative Analysis, Network Analysis, Spatial Analysis, Stylometry, Textual Analysis, Topic Modeling, and Visualization, and then introduces some special considerations regarding programming for the Humanities, including the place of money (a.k.a. funding) in the process, the difficulty researchers face when they attempt to quantify the Human, qualitative analysis as an alternative and its usefulness for DH projects, and the discussion about whether DH researchers should focus more on the proper use of the tools that already exist, over creating their own one-off tools.

Keywords Humanities data · Digital-humanities approaches · Digital archiving
Network analysis · Spatial analysis · Stylometry · Textual analysis · Topic modeling · Visualization

3.1 Digital Humanities Programming

The introductory programming course is first, and foremost, taught to help students learn how to design and write computational solutions for given problems. In the teaching of a programming course for the Digital Humanities (DH), we should not interfere with that goal. The inclusion of DH assignments must provide learning experiences which support this unalterable path. Luckily, trying to find computational solutions to help students also learn more about research methods which study humans and human culture is not much different from finding computational solutions for problems in any other domain.

In Chap. 1, I provided a brief overview of the DH field and the introductory programming/programming for the humanities course I developed after my college initiated a DH minor program. In this chapter, I discuss the special considerations

B. Kokensparger, *Guide to Programming for the Digital Humanities*, SpringerBriefs in Computer Science, https://doi.org/10.1007/978-3-319-99115-3_3

that introductory programming instructors must keep in mind when using DH assignments to help all students practice fundamental programming skills, while also allowing DH students to become familiar with basic DH computational methods.

3.2 Popular Analytical Approaches to Digital Humanities

There are a large number of analytical approaches to DH research, perhaps as varied as the research projects, themselves. However, having done a survey of current DH literature and conference proceedings, there are a few approaches that seem to recur frequently [1]. The DiRT database summarizes and offers tools for a number of DH analytical approaches [2] as well. NEH workshops also provide a good overview of general approaches to DH, including the curriculum for the EMDA 2015 institute [3].

These approaches can be grouped and summarized (with my definitions and descriptions of their general approaches):

- *Digital Archiving and Editing*—a systematic method of obtaining, digitizing, and making artifacts accessible in an enriched environment—this method is introduced in the Code Transformation assignment in Chap. 8.
- *General Quantitative Analysis*—a systematic method of measuring and quantifying artifacts of human production using statistical methods, such as frequency analysis, correlations, and regressions—this method is introduced in the Change over Time assignment in Chap. 5.
- *Network Analysis* (includes Social Network Analysis)—a systematic method of observing and analyzing connections between entities, be they people or things—this method is introduced in the Social Network Analysis assignment in Chap. 10.
- *Spatial Analysis* (includes Geospatial Analysis)—a systematic method of analyzing an artifact in reference to its location, both in relation to other physical characteristics of the Earth and in relation to other artifacts. I have not yet developed an assignment introducing this method to novice programming students.
- *Stylometry*—a systematic method of analyzing an artifact (often text) to classify it in terms of style—often used for author attribution projects—this method is introduced in the Art Stylometry assignment in Chap. 9.
- *Textual Analysis* (includes Sentiment Analysis and Corpus Analytics)—a systematic method of measuring attributes of one or more texts—often used in applied linguistics—this method is introduced in the Frequencies and Stop Words assignment in Chap. 7.
- *Topic Modeling*—a systematic method of discovering summarization topics in given artifacts and determining how they relate to other topics in the same or other artifacts. I have not yet developed an assignment introducing this method to novice programming students.

- *Visualization*—a systematic method of displaying data using visual elements to convey findings and results, and as a method of further analysis—this method is introduced in the Visualizing Change over Time assignment in Chap. 6.

Though some DH researchers may argue about my inclusion of one analytical method and exclusion of another, or perhaps may disagree with the name that I use to characterize a specific method, or my definition of a method, I suspect that most DH experts would agree that these approaches are a good place to start, and by using them no "evil" would be done to the students or to the DH field. These topics are corroborated in a taxonomic survey that I published a few months before this book was published [4].

The least common denominator among all of these DH computational methods is the bringing together of the Human and the data which are collected about the Human in reference to some finite time and place, for the reason of learning more about the Human within that context. With this charge as a core activity, the aim of the DH assignments in the course is to help students begin their personal journeys to find the data in the Human, and thereby find the Human in the data.

Inherent within these journeys are some special considerations that are peculiar to DH research.

3.3 Special Considerations in Digital Humanities

Unlike business, where the point of the analytical effort is to determine better ways to increase revenue and decrease expenses (i.e., make money), the point of the effort in DH is to find the Human and to learn more about the Human in the process. As such, DH researchers look through an entirely different lens. What is the measure of success for a DH researcher? There is no monetary value that can be placed on learning about the Human, especially if there is no extension of that knowledge towards motivating the Human to make a favorable response to stimuli.

That is the first among the special considerations regarding DH research: money is external to the process. DH research is often supported through grants, special arrangements, benevolence, and donated in-kind services. No one has ever gotten rich studying the Human, except in cases where the output has become largely commercialized. In an introductory programming classroom, discussion of the DH assignments should be coupled with a genuine love of research for the purpose of adding to knowledge about the human condition and humanity's cultural and social past. If you have no appreciation of this pursuit, you may not be the best person to teach a programming for the humanities course.

A second special consideration, regarding DH research, is that humans are really difficult to quantify. Humans cannot be isolated from their environments, and the people with whom they interact during a research study. Even the most well-planned research study on humans must be done in a place, with real people, so there is always the possibility that biases will occur in the subjects that will skew the results.

There are simply too many variables, so if we treat humans, we're never really that sure that our quantitative analysis is effective. In an introductory programming classroom, you should introduce each of the DH assignments with the caveat that the given computational method is not perfect; and even the most complex DH tool attempts to quantify or illustrate phenomena in ways that produce a certain degree of data loss. It is a fundamental truth that a visualization, for example, cannot illustrate every dimension of data discovered around even the simplest of projects. There is a certain amount of data loss, either through deliberate selection and reduction, or through accidental collection and analysis errors. Choices must be made, and whenever and wherever there is a choice, there is data loss.

A third special consideration of DH research, is that qualitative analysis works well with humans, but does not directly employ the computational power of computer technology. There are some who say, I'm sure, that they have created qualitative analysis engines that rival a human's manual processes, and that deep learning (through neural networks and genetic algorithms, for example) has and will continue to make inroads into the automation of this process [5]. But at this point of time, the bulk of DH work is done in a quantitative arena.

A fourth special consideration surrounds the idea that DH researchers should focus more on the tools that already exist, and how to use them properly. From this argument, a programming for the humanities course should be more focused on the expert use of existing tools, such as Gephi (for topic modeling) and ArcGIS (for geospatial analysis). I once spoke informally with a world-renowned DH researcher who did not think it worthwhile to teach students how to write code to do visualization, for example. "There are plenty of visualization tools out there," the researcher said. "Just teach your students how to use them properly." I personally think that any given tool adequately answers the questions of the person who programmed it; any other researchers who use it blindly, do so at their own risk. Working only with existing tools may be a good start, but any real DH programming problem usually involves some modification of the tool in terms of recompiling source code or creating libraries and add-ins, or sometimes even creating an entirely new tool.

Teaching introductory programming students the basic programming approaches to DH computational methods arms them with not only the knowledge of what to do with a given tool, but at least a little bit of knowledge about why to do it and a glimpse into a tool's limitations, either published or hidden. This approach gives DH students a realistic view of the computer's role in their future research, both its beauty marks and its moles. The better informed they are, the better users they will be, and ultimately the better research they will do in the field.

Now that we set the context regarding the teaching of DH methods in an introductory programming course, we can now focus more on the assignments themselves, starting with a basic overview in the next chapter.

References

1. CUNY Academic Commons Wiki Archive. The City University of New York. https://wiki.commons.gc.cuny.edu/tools__methods/ (2018). Accessed 19 June 2018
2. DiRT Digital Research Tools. Andrew W. Mellon Foundation. http://dirtdirectory.org/ (2018). Accessed 19 June 2018
3. Early Modern Digital Agendas 2015 Curriculum. The Folger Library. https://folgerpedia.folger.edu/EMDA2015_Curriculum (2015). Accessed 19 June 2018
4. Kokensparger, B.: What ought to be taught?: topical analysis for teaching future digital humanities researchers in a CS1 course. J. Comput. Sci. Coll. **33**(5), 165–171 (2018)
5. Svensson, P.: Humanities computing as digital humanities. In: Defining Digital Humanities, pp. 175–202. Routledge (2016)

Chapter 4
Introduction to the Digital Humanities Assignments

Abstract The six Digital Humanities assignments provided in this book provide support for students to learn all of the basic programming concepts covered in any introductory programming course, while simultaneously acquiring experience in appropriate DH analytical areas. In this chapter, some general tips to address issues, which include determining what level of the assignments is best for a specific set of students, the iterative approach (introducing difficult structures and concepts early and discussing them fully later in the semester), and using scaffolding as a desired pedagogical method, are presented as a focus on managing the assignments throughout the semester. This chapter also discusses the assignment chapter conventions, including the basic assignment descriptions and support materials provided in each assignment chapter.

Keywords Introductory programming · Difficulty level · Iterative teaching Educational scaffolding

4.1 Finding the Balance Between CS and DH

When I was first given this task of creating Digital Humanities (DH) assignments for an introductory programming course, I was faced with a serious dilemma: On one hand, I wanted to make sure that the DH students received a good introduction to some of the basic methods of DH research, such as social network analysis, stylometry, textual analysis, and visualization. But on the other hand, I was merely teaching an introductory programming course, so I could not expect all of the students to do professional-grade work, or to write analytical code at the level of functionality expected of advanced computer science practitioners.

My solution was to try to strike a middle ground between the two ends of the spectrum—to give all students a solid introduction to DH analytical methods while keeping the course accessible for novice programmers, using assignments that employ the strengths of the Python language. So the average student in the course should be

B. Kokensparger, *Guide to Programming for the Digital Humanities*, SpringerBriefs in Computer Science, https://doi.org/10.1007/978-3-319-99115-3_4

able to do a good job of programming the assignment, and should also be able to learn something about how programming for humanities research is done.

What is particularly difficult about this task is that it attempts to please a wide range of students. At risk are the computer science (CS) students who only want to learn how to program, and have no interest in the humanities, as well as the DH students who have a great interest in the humanities, but have little or no desire to learn computer programming. The great benefit of having these DH programming assignments is that they are unusual. If you browse the more popular introductory programming in Python textbooks, most of the assignments are math-based, science-based, or business-based. The web is also filled with such assignments. So CS students with no interest in the humanities have a whole web of assignments to use for practice, if they wish to avoid the humanities. Also, DH students who have little interest in learning programming might at least be brought along with assignment topics that are of more interest to them.

In summary, the six assignments that are provided in this book are all designed to do a good job of splitting the difference between offering real problems with real issues that require real solutions, yet are also accessible to real novice students (if completed in order).

4.2 General Assignment Management Tips

In each of the following assignment chapters, there is an *Assignment Management Techniques and Issues* section, that provides suggestions and best practices for implementing the assignment in your classroom. These suggestions are focused specifically on that assignment, so there is little transference between chapters. However, there are some assignment management tips that apply generally to all of the assignments. It might be helpful to cover some of them here.

The first two assignments, among the six, are specifically designed to provide gentle low-risk experiences for the novice programmer. As instructors, we do not want to scare the novice students away by pushing them into the deep waters of solving complex programming problems before they have gained their confidence, as well as some mastery over the Python syntax.

If you are using the first two assignments with intermediate or advanced students, you may need to employ more challenging versions of those assignments, to avoid boring them with problems that are too easy. In each chapter, I have provided suggestions for altering the given assignments for novice, intermediate, and advanced students, as well as suggestions for more homogenous classrooms of secondary school students and DH (non-CS) students. As you grade student submissions for the first two assignments, you will get a feel for which suggestions to implement that are right for the group of students with whom you are working. Your approach may even change from semester to semester.

It also might be a good strategy to simply ask the students which level of assignment they would like to do, and let them seek out their own challenges. This might require a longer grading process on your end, but may also increase student learning on theirs. Or some instructors may want to offer an "A assignment" out of the advanced suggestions, a "B assignment" out of the intermediate suggestions, and a "C assignment" out of the novice suggestions (which are proposed in each chapter as the "base" assignment), for which the assignment description is written. Again, this might require more grading on your part, but could be a big step in helping students take control of their own learning.

The assignments in this book use an iterative approach: Students are gently introduced early in the semester to structures, like loops and branches, that they will learn about in detail later in the semester. In these early introductions, students are given code to use and instructors are urged to provide demonstrations and talk through what is happening. Then, later in the semester, when students are officially introduced to the structures in detail, they will hopefully understand them more fully due to their prior experience with them. This follows the approach used by Dr. John Zelle in *Python Programming: An Introduction to Computer Science* [1]. If you use a different approach, one where you do not introduce the more difficult structures for novice students until they are fully discussed in class, then you may want to either provide a lot of code for your students to just use in the first two or three assignments (early in the semester), or change the requirements of the assignments to eliminate any use of structures that will not be introduced until later. This latter approach runs a great risk, though, that students will not be able to do anything "worth doing" until the second half of the semester.

I also suggest that instructors use the pedagogical approach of scaffolding to present these assignments. Early in the semester, there should be a great deal of in-class interaction, to ensure that all students have the support they need to successfully produce an appropriate (or approximate) solution. Then, as the semester goes on, you should try to back out of the process to allow students to find and utilize their own resources as much as possible. This approach is considered counter-intuitive to some CS educators, who question why students should get more support on the easy assignments and less on the difficult ones. But as it is our job to not only teach students how to program, but also to teach them how to learn how to program, it is doing them a favor in the classroom to encourage them eventually to lean on their own internal resources and, where needed, identify appropriate external resources.

Assignment-specific suggestions and best practices are provided in each assignment chapter (Chaps. 5–10), in addition to a number of other helpful features, as previewed below.

4.3 Assignment Chapter Features

There are six assignment chapters, each of which describes and provides materials and support for one of the DH assignments. It might be helpful here to discuss what features are provided for each of the assignments, in the various chapter sections. Each assignment chapter includes:

The Assignment Description—This section provides verbatim description and directions, suitable for copy-and-paste, of the material that I give to my students at project assignment time. These descriptions have been developed over the years to facilitate student understanding and set students up for success in the assignment. The assignment description in this section is intended to be given to students as is, or modified based on your own educational contexts. The material in the other sections of each assignment chapter is not written in a way to be given directly to students.

Assignment Files and External Resources—This section lists files and other resources the students must have to complete the assignment, as well as some alternative links for supplemental material that may be helpful to you. Links are provided to the files themselves on a server, provided by Springer Publishing, which should be available for a duration consistent with the feasible use of this book.

Skills Utilized in this Assignment—This section provides a list of both Python programming skills and digital humanities skills that the students will practice while completing assignment as described in the chapter. This assumes that students will also have practiced all of the skills listed in previous assignments, so if you change the sequence of the assignments, you might take heed that students may not be well-prepared for that chapter's assignment in terms of the listed skills from prior assignments.

Assignment Management Techniques and Issues—This section provides comments and suggestions regarding issues of which every instructor should be aware when managing the current chapter's assignment. There are issues that have arisen every semester, and which I expect to arise again in the classroom next time that I offer that chapter's assignment. There are also some "gotchas" and learning opportunities that come from them, all of which are briefly discussed. These management techniques are not intended to be given to the students. For example, in some assignments I may give "step back" advice for when students are struggling mightily with the assignment. These things sometimes happen. It would not be good for students to see this suggestion in their assignment materials—it may motivate them to "struggle" mightily when they otherwise may not have done so.

Atomic Code for the Assignment—This section provides a coding assignment that should be done in class, which will give your students the basic experiences that they need to be successful in completing the actual DH assignment. In other words, if they can complete the atomic code assignment, they should be able to complete the DH assignment. This also gives you the opportunity to show, share, and explain the atomic code, so that students with less experience will have working code in hand that they may be able to rework to some degree for the assignment. The subject domains of these assignments are not generally from the DH field, which provides students both with some variety in subject domains, but also may help them see how strategies and patterns are often similar in different subject domains. In other cases, especially among the later assignments, the atomic code is simply the starter code for the assignment, itself.

Expected Output from Student Work—This section shows what the students' work should produce, more or less, based either on real student work that has been submitted in the past (with the student's permission, of course), or output from my own proof program. The information in this section could be helpful when formulating a grading strategy for the assignment, or when creating your own proof code for the assignment, or—in rare cases—when showing students what should be provided by their code. In the case of the latter, in the early assignments, when students are asked to compile data manually, providing the output code will result in "giving away" that part of the assignment, so you want to be careful about providing this expected outcome information to students in your class.

Assignment Variations—This section provides suggestions to vary the assignment based on the skill level and age of students that you might have in your classroom. These variations cover:

For Novice Students (taking an Introductory Programming Course)—introductory programming students with very little—perhaps no—coding experience. This is the group of students for which the assignments were created, in their original forms. If you have a classroom with mostly novice students, then you will most likely need to make very few changes to the assignment descriptions and directions provided.

For Intermediate Students (taking a Python Programming Course)—students who have already completed an introductory programming course, and are perhaps enrolled in a Programming in Python course. These students already have familiarity with general programming structures (like variables, loops, etc.) but may have little to no familiarity with Python programming syntax. Assignment variations for this group normally assume that structures normally taught later in an introductory programming semester—like loops—are available to the student in a higher-than-introductory-familiarity level early in the semester.

For Advanced Students (taking a Software Engineering or Capstone Course)—students who have a lot of coding experience and already know how to solve computational problems in Python. Assignment variations for this group are focused more on a software engineering or data structures approach, emphasizing bigger projects extended from each assignment's core idea.

For Secondary School Students—students who are younger students, who are assumed to have a high level of motivation to learn through the assignment but who have a low level of academic maturity to handle the tedium of programming. Variations on the assignments for this group include suggested "unplugged" activities and classroom pedagogy that is more active and tactile. As I do not have an opportunity to teach secondary school students, I have not tested these suggestions. Note, motivated high school students are more like novice college-level introductory programming students, and should probably be treated as such. There may also be secondary school students in your classroom who are prepared to handle the intermediate or advanced suggestions for the assignments. As always, your judgement of your students' capabilities is usually the best data for making changes to the base assignments.

For Digital Humanities Students—students who wish to learn to program but are homogenous in their motivation to work in DH. These students need to learn all of the basic constructs of programming, but variations for these students include ways to spend more time and provide more direct experience in DH activities. This is somewhat analogous to a swimming class—if the instructor knows that every student in the class eventually wants to learn to SCUBA, more instruction in communication and underwater survival methods might be provided earlier in the curriculum.

The next six chapters include the descriptions and support material for each of the DH assignments, themselves.

Reference

1. Zelle, J.: Python Programming: An Introduction to Computer Science, 3rd edn. Franklin, Beadle, Portland, Oregon (2016)

Chapter 5
Change Over Time: Burials in an Historic Cemetery

Abstract The first of the six Digital Humanities assignments, Change Over Time, is presented in this chapter, beginning with an introduction to the data source (Prospect Hill Cemetery, in Omaha, Nebraska). The assignment description (written in a form and with a point-of-view suitable to be copied and pasted into materials given directly to students), required support files and resources, skills utilized in the assignment (creating new files in the IDE, and simple input, output, and expression statements), assignment management techniques and issues (mostly involved with keeping students from jumping ahead on the assignment), atomic code for the assignment (a GPA calculator exercise), expected output from student submitted work in this assignment, and variations for a number of student skill levels are provided.

Keywords Introductory programming · Programming assignment
Python assignment · Historical analysis · Cemetery records · GPA calculator

5.1 The Phenomenon of Change

Take a moment to think of a specific classroom where you teach. Visualize the layout of the classroom. Is it arranged with individual desks or are there movable tables and chairs? Are there computers on the tables, or electric outlets? Think about what you use to teach your courses in that space. Do you use a computer? Do you project what's on your computer to a screen? Do you use a slide deck (like Microsoft PowerPoint™)? Do you use a white board with erasable markers? Most college instructors generally have the same general environment where they teach.

Now take a moment to think back to a specific classroom where you learned as an elementary school student. If you are anywhere over 30 years old, there are most likely some striking differences in your vision of a classroom where you learned, compared to the vision of where you currently teach.

Electronic supplementary material The online version of this chapter (https://doi.org/10.1007/978-3-319-99115-3_5) contains supplementary material, which is available to authorized users.

B. Kokensparger, *Guide to Programming for the Digital Humanities*, SpringerBriefs in Computer Science, https://doi.org/10.1007/978-3-319-99115-3_5

In my case, I currently teach in a classroom that allows me to project my computer onto a screen at the front of the classroom. I use a slide deck, and we have tables in the middle of the room, where most of my students open and use their laptop computers. Around the edges of the classrooms are tables holding desktop computers, for students who do not have laptops. I draw on a white board with erasable markers.

However, as an elementary school student, I sat in straight rows in hard wooden desks. At the front of the room were green chalk boards, with white chalk and erasers that had to be clapped together at the end of the day. When the teacher wanted to show us content, she fired up the filmstrip machine with a cassette recorder, which beeped every time the image had to be advanced. Some teachers also drew on acetate sheets on an "overhead" projector, or fed film through a film projector's workings to show moving images.

If educational technology historians took snapshots of the two learning spaces, they would see vast differences between the technologies employed. They would ask questions about why, in both cases, those technologies were employed, what exactly they provided in terms of educational value, and what conditions led to the adoption and employment of the new technology. In other words, these historians would be interested in the concept of *change over time*.

Analyzing change over time is a staple of historical research. Therefore, there are a number of digital humanities (DH) tools that help historians track, discover, analyze, and communicate findings regarding changes over time. These tools are easily found through simple web searches.

As we will note in all of these assignments, computers are exceptional at carrying out repetitive tasks and calculating expressions more quickly and (if properly programmed) with fewer errors than their human counterpoints performing manual operations. However, sometimes a case can be made to employ manual operations over computer-assisted ones. This DH assignment shows us the benefits of both sides, and provides an experience that should engage novice programming students right where they are in the beginning week of their introductory programming course.

For this DH assignment, we will focus on developing programming code to solve one problem: evaluating change over time in burial permit records in an historic cemetery.

The Prospect Hill Cemetery is an historical cemetery just a few blocks away from my university (Creighton University), in Omaha, Nebraska. It has records of burials as far back as 1855, but there most likely are burials that existed well before that date. Burials, removals, reburials, and payments on account (among other matters) are recorded in a document called a *burial permit*. I am working with students and alumni of our institution to digitize all 13,500 of the burial permits, which must be cropped from scans, uploaded to a server, and then digitally hand-entered from the volunteer's interpretation of hand-writing presented on each burial permit.

Though it is a laborious process, over 6000 permits have been entered so far, and this assignment uses a selected number of fields from a small subset (1368) of the burial permits, ranging in interment date years from 1872 through 1878.

For their first DH assignment, students will create a change over time Python program to analyze data from these burial permits and submit the Python solution file.

5.2 The Change Over Time Assignment

The world is ever-changing, and has been through history. Any photograph taken from the exact same angle with the same camera a day later will show changes of one type or another. Historians track these changes using a method called *change over time*. That is what you will be doing for your first program for this DH assignment.

In the files for this course is an Excel spreadsheet called PHCPartialBurial-Data.xlsx. It contains 1368 Burial Permit records, all of which are sorted by interment (burial) date. These burials occurred between the years 1872–1878. Figure 5.1 shows a selected excerpt from the provided file.

Your task for this assignment is to combine and modify the programming code provided in the atomic code below to create a Python script that takes as input the number of burials (i.e., number of burial permits reported) for each year and provides the yearly average of the number of permits issued over the entire 7 years.

Input: User inputs the year identifier (e.g., 1872), and the number of burial permits for that year. Note the permit numbers in the spreadsheet have letters appended to the beginning and end of the field, which should be omitted.

Process: As each year's data are entered, the total number of burial permits for that year should be added to the *grandTotal* variable. After all of the years have been entered, the average number of permits should be calculated (*grandTotal*/*numYears*) and stored in a separate variable (*avgBurials*).

	A	B	C	D	E	F
1	BurialPermit	BurialDate	DecLast	DecFirst	DecAge	LotNum
2	BP01128a	Jan 2 1872	Bruner	Alice	3 yrs & 3 days	A938
3	BP01129a	Jan 3 1872	Rankin	S	45 years	A831
4	BP01130a	Jany 5 1872	Bowman	Ann	50 years	A938
5	BP01131a	Jany 7 1872	None	None	5 weeks	A826
6	BP01132a	Jany 10 1872	Judson	Philo	83 years	A831
7	BP01133a	Jan 20 1872	Moore	Frankie	7 mos	A826
8	BP01134a	Jan 21 1872	Windheim	August	34 years	A1005
9	BP01135a	Jany 24 1872	Elliott	Flora	4 years & 5	A912
10	BP01136a	Jany 24 1872	None	None	4 days	A826
11	BP01137a	Jan 25 1872	Smith	Nelli	6 weeks	A826
12	BP01138a	Jan 27 1872	Hausen	Anna	2 yrs & 2 mos	A826
13	BP01139a	Feby 1 1872	Chatterton	Ella	14 years	A840
14	BP01140a	Feby 1 1872	Parker	Charles	about 43	A797
15	BP01141a	Feby 8 1872	Lamb	A	18 years & 7	A199
16	BP01142a	Feby 12 1872	McCheane	Mary	43 years	A811
17	BP01143a	Feby 13 1872	None	None	1 day	A826
18	BP01144a	Feby 13 1872	Parker	Mary	about 1 year	A797
19	BP01145a	Feby 15 1872	McLean	Archibald	about 34 years	A840
20	BP01146a	Feby 17 1872	Phillips	William	about 22 years	A844
21	BP01147a	February 18 1872	Sievers	Mary	3 weeks	A828

Fig. 5.1 Sample excerpt of the data file for the change over time assignment

Output: Besides printing out the year and the number of permits per year, the overall average number of permits per year over the 7-year period (*avgBurials*) entered should be printed out to the display as well.

For this program, put the data to input within comments near the program header, and turn in your solution code as a [lastname]Change.py file.

Though pseudocode will not be provided for all of our assignments, it is helpful in this case to provide you with pseudocode so that you know you are on the right track, and to keep you from complicating the problem unnecessarily (like trying to use structures, such as loops, that you have not learned yet).

Pseudocode (for beginning students):

Create and initialize variables (*numYears*, *inputYear*, *inputBurials*, *grandTotal*, *avg-Burials*)
Have user input the **first** year (into *inputYear*) and number of burials for that year (into *inputBurials*).
Add number of burials for that year (*inputBurials*) to *grandTotal*.
Print the year (*inputYear*) and number of burials (*inputBurials*) for that year.
Have user input the **second** year (into *inputYear*) and number of burials for that year (into *inputBurials*).
Add number of burials for that year (*inputBurials*) to *grandTotal*.
Print the year (*inputYear*) and number of burials (*inputBurials*) for that year.
Have user input the **third** year (into *inputYear*) and number of burials for that year (into *inputBurials*).
Add number of burials for that year (*inputBurials*) to *grandTotal*.
Print the year (*inputYear*) and number of burials (*inputBurials*) for that year.
Have user input the **fourth** year (into *inputYear*) and number of burials for that year (into *inputBurials*).
Add number of burials for that year (*inputBurials*) to *grandTotal*.
Print the year (*inputYear*) and number of burials (*inputBurials*) for that year.
Have user input the **fifth** year (into *inputYear*) and number of burials for that year (into *inputBurials*).
Add number of burials for that year (*inputBurials*) to *grandTotal*.
Print the year (*inputYear*) and number of burials (*inputBurials*) for that year.
Have user input the **sixth** year (into *inputYear*) and number of burials for that year (into *inputBurials*).
Add number of burials for that year (*inputBurials*) to *grandTotal*.
Print the year (*inputYear*) and number of burials (*inputBurials*) for that year.
Have user input the **seventh** year (into *inputYear*) and number of burials for that year (into *inputBurials*).
Add number of burials for that year (*inputBurials*) to *grandTotal*.
Print the year (*inputYear*) and number of burials (*inputBurials*) for that year.
Calculate *avgBurials* by dividing *grandTotal* by *numYears*.
Print out *numYears* and *avgBurials*.

Note there is a lot of repetition in the pseudocode. Concentrate on each individual line of the pseudocode, writing a line of Python code in your file that corresponds with each line of pseudocode. Later you will learn loops and other structures that will help reduce the lines of code needed to solve this problem. For now, be patient and don't worry about the repetition. Remember, copy-and-paste is your friend!

5.3 Assignment Files and External Resources

This file is needed to complete this assignment: PHCPartialBurialData.xlsx

5.4 Skills Utilized in This Assignment

This assignment is designed to give the novice student some autonomy in terms of writing code. It's not meant to teach a large number of programming concepts, per se, except for opening up the Integrated Development Environment (IDE), creating new files in the IDE, and performing simple input, output, and expression statements.

Programming Skills Utilized:

- Opening and managing the integrated development environment (IDE)
- Creating a new Python file
- Simple input statements, with type-casting
- Creating an "accumulator" variable and using it with multiple value inputs
- Simple expressions
- Simple output statements (*print* method)
- Value of offline/manual operations preliminary to writing code.

Digital Humanities Skills Utilized:

- Exposure to real-life recordkeeping in the late 19th Century
- Simple data analysis/data reduction
- Data interpretation
- Change over time as a historical concept
- Cemetery records as data sources.

5.5 Assignment Management Techniques and Issues

This assignment should be given to the students early in the course semester or quarter, probably the first week of a normal 16-week college semester for introductory-level students.

Some students will want to use Python code to input the spreadsheet as a CSV file and do the counting work for them. This is laudable, and a practice that might be expected from advanced programmers. But the students in your class are likely not advanced programmers. Reinforce the idea that the manual processes (like counting with tic marks) and semi-manual processes (like dragging over cells of a spreadsheet to count rows for them) are important. Even though libraries have been created by the Python community for just about any task imaginable, it is too early to introduce specialized libraries to your introductory students—they need to learn how to solve problems for themselves before they use other users' solutions.

Even though it may seem counter-intuitive, I suggest not giving students in-class time to work on this assignment, as I believe it is important for students to take ownership over their individual problem-solving efforts early in the semester. This will force them to set up their "at home" programming environment, and begin establishing a workflow and norms for successful programming code completion.

Some novice students attempt to use a loop (which hasn't been covered in class yet, despite the fact that the textbook I use gives a brief introduction to the topic in the first chapter) to solve this problem. I usually discourage students from using loops for this assignment, because the copy and paste approach to solving this problem is still a valuable skill to learn as a novice programmer. That's why pseudocode is provided for novice students—to show them that it is okay for them to employ repetitive code instead of looping structures at this point. Later, they will learn the structures to reduce the repetition to a few statements in a well-crafted loop. Of course, intermediate and advanced students should be required to use looping structures and any other Python conventions suitable for the task.

Some students will also attempt to use different variables for each year's input, which complicates the program unnecessarily. If they follow the pseudocode for this assignment, they will be encouraged to reuse the same variables for input of each year (*inputYear*) and the burials for that year (*inputBurials*). That is sufficient for solving this problem. Though novice students often err on the side of creating too few variables for the task, students who create new variables for each of the seven years should be counselled to only use the original two.

Some novice students may also try to use a list (array) to store all seven years' data. This is a good opportunity to discuss the assessment of how long inputted data needs to be retained to solve the problem. In the case of this assignment, it is sufficient to keep the data until that individual year's data has been printed and the total burials for that year added to the accumulator variable (*grandTotal*). After those tasks have been completed, for this problem, each year's data is no longer needed, and should not be retained by the program.

5.6 Atomic Code for This Assignment

Among the things that novice students should learn in completing this assignment, they should become familiar with the integrated development environment (IDE), including how to open a new file for code. Figure 5.2 shows how to do this in the IDLE shell.

Computer programs are able to complete tasks that are manually tedious but without human error. One of these tasks is performing a grade point average calculation for a semester.

Your challenge module will be to create a Python script that takes as input a course identifier (like CSC221), the letter grade received in the course (A, B+, B, C+, C, D, or F), and the number of points associated with the letter grade. The script should calculate the resulting points (3 hours of a course with a grade of B+ would yield 3 * 3.5 or 10.5 calculated points) and keep track (in an accumulator variable) of the sum of all of the calculated points and the total semester hours. The program should be set to collect grades for 5 courses of 3 semester hours each.

Input: course identifier, letter grade, number of points for that grade (as shown in the distribution below, which should be displayed for the user's convenience at input time).

Process: Calculate the total points for the course, add total points to an accumulator variable.

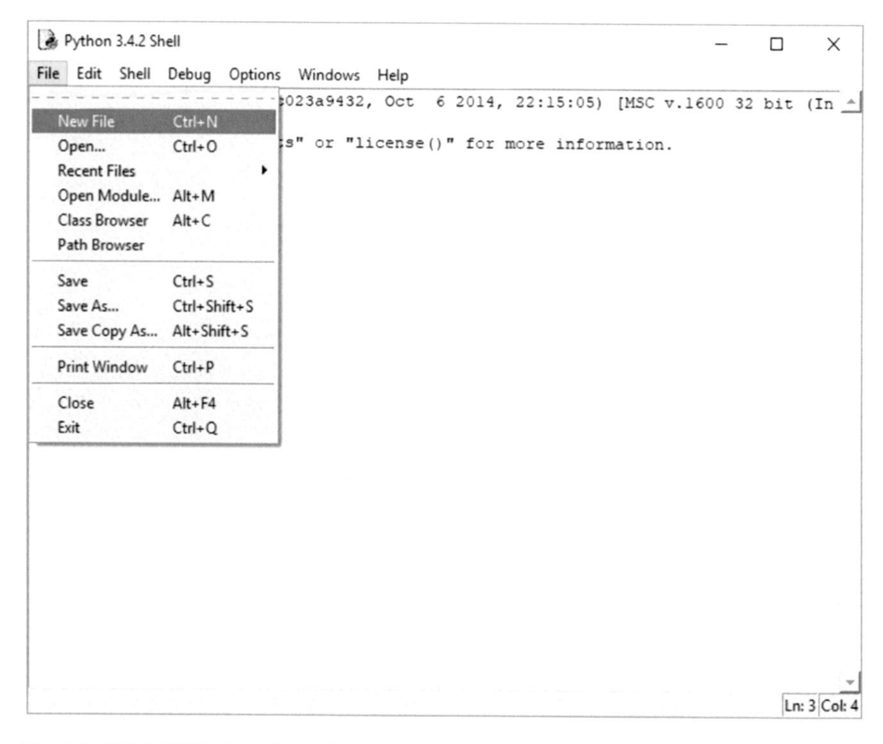

Fig. 5.2 IDLE (IDE) view of creating a new file

Output: After the user enters each line, output the data for that line. Then, after data for all 5 courses have been entered, output the grade point average for that semester.

Points are distributed as such:

A 4 points
B+ 3.5 points
B 3 points
C+ 2.5 points
C 2 points
D 1 point
F 0 points.

5.7 Expected Output from Student Work

Students may be given some leeway in fulfilling the requirements for this assignment. However, Fig. 5.3 shows an example of the type of output that student work may produce.

5.8 Assignment Variations

5.8.1 For Novice Students (Taking an Introductory Programming Course)

This is the first assignment for my introductory programming students, so I expect very little in terms of programming style and knowledge. Students can pretty much do the program however they want to do it, and though the output is often a bit ugly it allows them to learn how to work with the IDE and how to get through debugging and creating a program that solves the problem.

5.8.2 For Intermediate Students (Taking a Python Programming Course)

For students who have adequate experience with programming, and are learning Python syntax, I urge them to use file input, loops, and formatted output and create a solution that is reasonably elegant. I also urge them to think about other data dimensions available in the data file, some of which might be analytically significant, and to provide output on those items as well.

```
Python 3.4.2 (v3.4.2:ab2c023a9432, Oct  6 2014, 22:15:05) [MSC v.1600 32 bit (
tel)] on win32
Type "copyright", "credits" or "license()" for more information.
>>> ================================ RESTART ================================
>>>
This program determines the average number
of people buried in Prospect Hill Cemetery
between 1872 and 1878.

Please enter the year -> 1872
Now enter the total burials for 1872 -> 179
For 1872 there were 179 burials.
Please enter the year -> 1873
Now enter the total burials for 1873 -> 191
For 1873 there were 191 burials.
Please enter the year -> 1874
Now enter the total burials for 1874 -> 193
For 1874 there were 193 burials.
Please enter the year -> 1875
Now enter the total burials for 1875 -> 168
For 1875 there were 168 burials.
Please enter the year -> 1876
Now enter the total burials for 1876 -> 192
For 1876 there were 192 burials.
Please enter the year -> 1877
Now enter the total burials for 1877 -> 184
For 1877 there were 184 burials.
Please enter the year -> 1878
Now enter the total burials for 1878 -> 261
For 1878 there were 261 burials.
The average burials by year over 7 years is: 195.42857142857142
>>> |
```

Fig. 5.3 Expected student output for the change over time assignment

5.8.3 *For Advanced Students (Taking a Software Engineering or Capstone Course)*

For students who have a lot of experience with programming, and are proficient with Python syntax, I urge them to employ all of the structures and suggestions in the intermediate version above, as well as to use existing Python libraries like the csv library to automate data input and aggregation. They should be required to aggregate other data dimensions (provided in the data file) to produce a fuller analysis of the data, perhaps using statistical methods.

5.8.4 *For Secondary School (Grades 7–12) Students*

As secondary school students are not that different than novice college-level students, they may trade youthful enthusiasm, present in an elective high school course or program, for a high school graduate's maturity and attainment of strategic academic skills. So, for this DH assignment, there are not many differences in the approach for this group from the novice college student group. However, middle school students may need to be introduced to spreadsheets and perhaps to the idea of death and burial of remains, and the record-keeping involved with this concept. Perhaps a visit to a nearby cemetery with a presentation by its designated record keeper might help the students. Another idea might be to substitute burial record data from a nearby historical cemetery in the place of the Prospect Hill Cemetery data provided with this book, and perhaps to involve the students, themselves, with data collection and spreadsheet production. This may provide a great opportunity for community-based service learning in your program.

5.8.5 For Digital Humanities Students

This is the first assignment for my introductory programming students, so the level of this programming assignment should be accessible and non-intimidating to any of your students. If the DH students have never coded before, they may need some additional coaching in building the repetitive nature of data entry, analysis, and output. A story such as Henny Penny (also known as Chicken Little) might elicit discussion on how the *grandTotal* grows as each year's data are added, culminating in a grand total by the end of the "journey" (i.e., the program). Like the Secondary School students listed above, a visit to a nearby historical cemetery, a discussion of burial permits and other recordkeeping tools, and perhaps experience with data collection to create their own spreadsheets of burial dates may be great additions to this assignment.

Chapter 6
Visualizing Change Over Time: Simple Visualization of the Burials in an Historic Cemetery

Abstract The second of the six Digital Humanities assignments, Visualizing Change Over time, is presented in this chapter, beginning with a brief introduction to visualization as an analytical method of great interest to humanities programmers. The assignment description (written in a form and with a point-of-view suitable to be copied and pasted into materials given directly to students), required support files and resources, skills utilized in the assignment (which include importing library files, instantiating and using objects, and programmatically managing color), assignment management techniques and issues (mostly involved with teaching students the value of manual processes, such as using pencil and paper to summarize data), atomic code for the assignment (a color change exercise), expected output from student submitted work in this assignment, and variations for a number of student skill levels are provided.

Keywords Introductory programming · Programming assignment · Python assignment · Data visualization · Cemetery records · Color management

6.1 A Language of Lines, Colors, and Textures

Even the trained eye benefits from a little help.

People, in general, are really good at detecting edges, and data visualization takes advantage of that characteristic by translating words and numbers into images with lines, colors, and textures. Lines, textures, colors, and representations of immediately-recognizable objects are used to communicate results. If visualization is done well, the data will be easier to comprehend, with more clarity and less chance for interpretation error. This is especially important if the results are geared towards an audience that is not trained in data interpretation.

Electronic supplementary material The online version of this chapter (https://doi.org/10.1007/978-3-319-99115-3_6) contains supplementary material, which is available to authorized users.

B. Kokensparger, *Guide to Programming for the Digital Humanities*, SpringerBriefs in Computer Science, https://doi.org/10.1007/978-3-319-99115-3_6

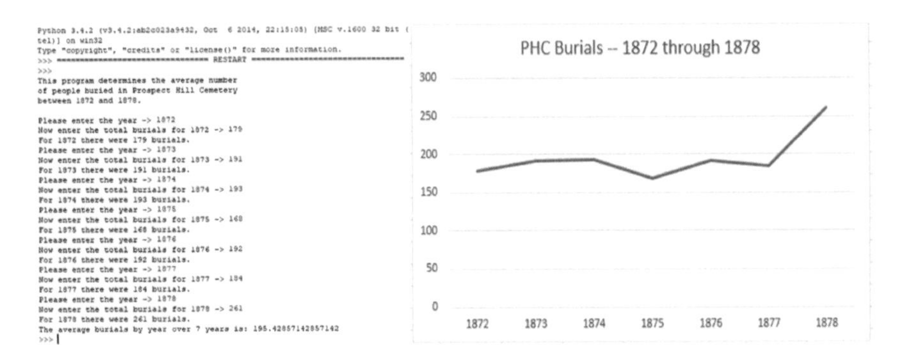

Fig. 6.1 Two views of the PHC burial frequencies per year from Chap. 5

Figure 6.1 shows the data from the Change Over Time assignment (from Chap. 5) in two different formats. The left side of the figure shows the year and the number of burials. The right side of the figure shows the data in what is probably the most common visualization method used today: the line chart. From viewing the line chart, there is an obvious upswing in number of burials in the final year. Is that upswing immediately apparent in the data on the left side of Fig. 6.1? To trained eyes it is, or even to the casual observer who takes the time to consider the data. But in either case, some comparative work has to be done by the viewer to interpret the data on the left and determine that the number of burials that occurred in the last year was quite a bit different from the numbers over the other years. Visualization, if done well, does this interpretive step for you.

Data Visualization is a staple of the digital humanities (DH) field, and is generally part of any DH research project. All DH projects have data of some sort (though the idea of what constitutes data might be stretched in some projects), and as we saw in the simple example above, well-done visualization enhances the communication of project results. Therefore, the concepts and practices of visualization are essential to any DH practitioner, and as computer science (CS) students will be working in any one of a number of disciplines, all of them collect, analyze, and interpret data, and then report results and conclusions.

Sometimes the data supporting *change over time* is simple and easy to analyze. The results may be preliminary inputs to bigger projects, or they may be so simple that a straightforward reporting of the data is sufficient. Most of the time, however, the results are more complex and need additional tools for analysis and reporting.

For this DH assignment, we will continue our focus (from Chap. 5) on analyzing *change over time* in burials from 1872 to 1878 in Prospect Hill Cemetery, by showing the changes graphically, using the graphics.py toolkit provided through Dr. John Zelle's excellent introductory programming textbook.[1]

[1] Dr. Zelle's excellent graphics library can be found at http://mcsp.wartburg.edu/zelle/python/, which provides a link to his graphics.py module and documentation). If you do not use this graphics library, there are other libraries you can use, or develop your own using the TKInter library (information available at https://wiki.python.org/moin/TkInter).

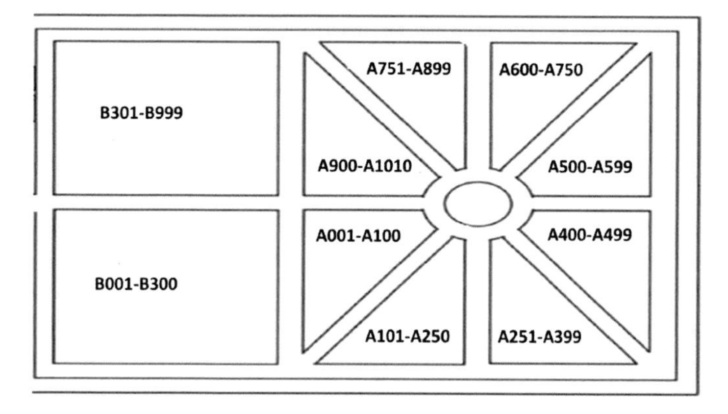

Fig. 6.2 Map of Prospect Hill Cemetery, in Omaha, Nebraska, showing lot number ranges for each geographic section of the main part and 1st addition of the cemetery

6.2 The Visualizing Change Over Time Assignment

Figure 6.2 shows a simple map of the Prospect Hill Cemetery, which reveals a number of *sections*, each of which contain a number of burial *lots*. Within each *lot* are a number of *parts*, which may contain burials and may be arranged in a number of ways according to the desires of the *lot* owners. The cemetery itself has a main area, a 1st Addition, and a 2nd Addition (the latter two of which were added after the initial limits of the cemetery were set). All of the lots are numbered in a consistent way, but lot numbers in the main section are duplicated in the 1st and 2nd additions, which complicates matters a bit.

In Fig. 6.2, the "main part" of the cemetery is depicted with lot numbers that begin with "A", whereas the 1st Addition of the cemetery is depicted with lot numbers that begin with "B". The 2nd Addition is not represented on the map, as there were no recorded burials there during our selected time period. The range of numbers shown in each section is the range of lot numbers in that section.[2]

In general, burials occurred during this time period in the Prospect Hill Cemetery in one of three ways:

Family members and friends of families who already owned a lot (generally families of means) were buried in that family's lot.

A deceased person from a family of means who was, perhaps, newly-arrived in the Omaha area, would be buried in a new lot purchased by the family.

[2]The range of actual lot numbers in the cemetery are not exactly as provided in the map. The lot numbers are roughly organized in the described manner, but were assigned informally and sometimes consecutive lot numbers extend over more than one geometric sector. For this assignment, the approximated ranges work well to give students a good, but simplified, experience in visualizing real-world data.

A deceased person with no family members in the area, or perhaps from a family with little means, were buried in lots that were not owned by specific families, so were thus opened to anyone who might need to be buried.

The PHCPartialBurialData.xlsx file (which was introduced in the previous chapter with the *Change Over Time* assignment) also contains a *LotNum* field, that provides the lot number where the burial occurred. Remember that an "A" at the beginning of the lot number denotes the "main" part of the cemetery, and a "B" at the beginning of the lot number denotes the 1st Addition of the cemetery.

Your assignment is to analyze the data from the PHCPartialBurialData.xlsx file and use the results of your analysis to build a visualization of how the burials were done in each of the represented sections over the entire seven-year period. This will require some pre-processing, that is, sorting the records by lot number and determining how many burials occurred in the range of lot numbers in each section over those seven years.

Once the number of burials in each section are computed manually, then you should import Dr. John Zelle's graphics.py file (which can be found at http://mcsp.wartburg.edu/zelle/python/), and create geometric objects (most likely the Polygon object for the triangles and the Rectangle object for the rectangles) to represent the sections of the cemetery in a visual re-creation of the map.

These objects should initially have a fill of white (color_rgb(255,255,255)), and then when the user clicks the mouse anywhere in the graphics window, should display a visualization of the aggregate (sum total) number of burials that occurred in each section of the cemetery during the 7-year period. One click should show all of the burials in the data file, with color shades that make sense in representing the given data.

In other words, your program should initially show the "map" of the cemetery with all areas having a default or "white" fill. When the user clicks the mouse, it should then show the cumulative burials after all seven years, so an accumulation of all of the burials that occurred over the seven years up to and including the year 1878. Note, in this basic assignment, the user only clicks once, and ALL of the data for ALL of the burials over ALL seven years shows as color fills in the map ONCE.

Figure 6.3 shows an example of how your final visualization might look (but with different data, so as to not give the solution away).

Intuitively, with most colors, the darker areas indicate sections that have more burials than the lighter areas. Red is, perhaps, an exception to that rule, as a "redder" red may be intuitively read as having more burials than a darker red. The choice of color and implementation of the various shades of the color are critical to the successful completion of this assignment. It is up to you to choose the color and the shades that you think best represent the total number of burials in a given section. Don't be quick to abandon shades of gray as a color palette. There is a saying in the viz biz: "Do it right in black and white."

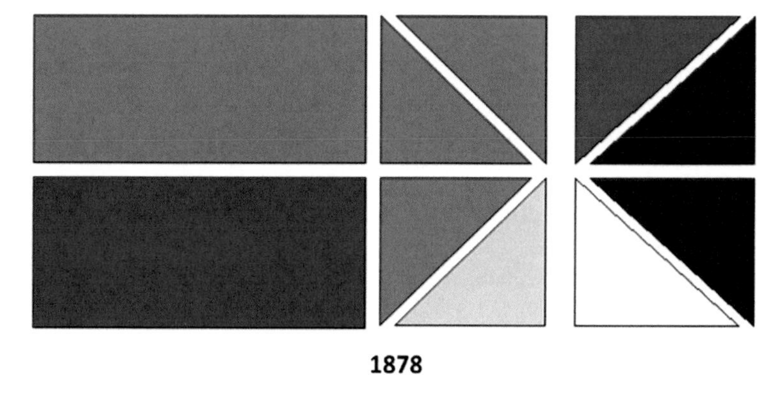

Fig. 6.3 PHC burial map visualization example using different data

It is also important to note the color ranges and how colors work in the World Wide Web. RGB is based on the colors red, green, and blue. So a value of color_rgb(255,0,255) will end up looking like purple (an even combination of red and blue) on the screen. Higher numbers (to a max of 255 in this system) generally mean the highest pixel intensity, or the "most lit" the pixel is in terms of that color. Consequently, a 0 number indicates a pixel "not lit" in that color at all. So color_rgb(0,0,0) is viewed on screen as black. Finally, if all of three of the numbers are of equal value (like color_rgb(128,128,128), then the color will be viewed on screen as a shade of gray. In this latter case, it would be a medium gray color.

Once you understand the way to make colors using color_rgb(int, int, int), you should play around with the color palette for a while to determine what rgb combinations best communicate the aggregate data that you wish to represent. Python statements with these values should be used in a call to the setFill(color) method of each of your Polygon and Rectangle objects that you created for this assignment.

Finally, please do not alter the graphics.py file in any way! The standard graphics.py file as available on the internet will be used to run your Python code during the grading process, and if you change the graphics.py file, those changes will not be available in my downloaded version of the file. Most likely this will break your program and result in a loss of points for the assignment.

When you are ready to submit your solution for this assignment, please only upload your <lastname> ChangeViz.py file with the proper header comments by the stated deadline.

6.3 Assignment Files and External Resources

These files are needed (or helpful) to complete the assignment:

Dr. Zelle's Graphics Library File
Dr. Zelle's Graphics Library Documentation
Student Work from Spring 2017
Prospect Hill Simplified Map
Prospect Hill Visualization Example
Assignment 2 Worksheet.

6.4 Skills Utilized in This Assignment

New Programming Skills Utilized:

- Importing a library file
- Instantiating objects in the library file
- Calling methods in instantiated objects
- Setting attributes in instantiated objects
- Using RGB values to change colors in objects
- Waiting for user mouse clicks to activate code.

Digital Humanities Skills Utilized:

- Simple geospatial analysis of data
- Representing a map with simple geometric shapes
- Determining a color palette for representation of data
- Visualization of analyzed data.

6.5 Assignment Management Techniques and Issues

This DH assignment should be given to your students early in the course period, probably the third or fourth week of a normal 16-week college semester for introductory-level students.

The students will ask about the availability of already aggregated data. I resist giving it to them—DH areas of study are filled with problems that are not cleanly handed to us at the beginning, and require some collection, cleaning, aggregation, and analysis. Especially for students, these data are rarely handed over already cleaned and organized—it's often the student's job to do this work. Additionally, students will gain familiarity with the data as they work with it, which cannot help but produce a better final project.

This is also a good opportunity to teach students how to read documentation when defining and using classes. Reading documentation and determining parameters and return values is a bit of an art, and the earlier students become familiar with the connection between library documentation and the usage of those objects, the better equipped they will be for the later course topics as well as when the time comes to write functions and class definitions for themselves.

The students should not attempt to try to duplicate the map exactly. A simple number of primitive objects, mostly Polygons (for triangles) and Rectangles will do fine. They should focus on these objects and their fill colors, not on additional lines, circles, etc. to try to re-create the whole map in detail.

As I indicated above in the assignment instructions, I always suggest that the students stick with one color palette. Multiple colors are difficult to deploy well in a range of values. Even though it's been done successfully in visualizations like weather maps (blue for snow, green for rain, pink for a wintry mix), these associations have been utilized for so long that they have become cultural icons that are known to most users.

More than most of the others, this assignment necessitates some paper work from the students. In addition to the manual operations of aggregating the data, making paper drawings of the objects and the changes to their fill colors helps them figure out what their color palette should be before writing a line of code. A helpful resource for this part of the assignment is the HTML Color Picker page at W3Schools (just search on "color picker" in any browser and it will be among the top hits).

For students who would like to attempt to use an expression to determine the fill color values for a map object, a straight linear expression may be the simplest solution, but will not be the best one. Instead, a few color shades (or "lightness" values in the HTML Color Picker), probably around 5, should be assigned to ranges of values that are similar. Some experimentation with colors should be taken on by the students to solve the problem, with the added benefit of allowing the students to gain familiarity with manipulating colors on an RGB scale.

Like the *Change Over Time* assignment (in Chap. 5), novice students may try to use lists and loops to avoid having a lot of lines of code in the module. It is imperative to explain to the students that they will learn these structures in future chapters, and will appreciate them more when they do!

6.6 Atomic Code for This Assignment

Four Seasons

Using the objects available in Dr. John Zelle's graphics.py module, use Python code to draw a picture of a tree in a graphics window. Then write code so that, with each click of the mouse, the tree foliage changes season. You can choose to start with any season, but your program should then progress through all four of the seasons. In addition to changing the season, draw text to the graphics window to identify the season, like "Winter" or "Autumn".

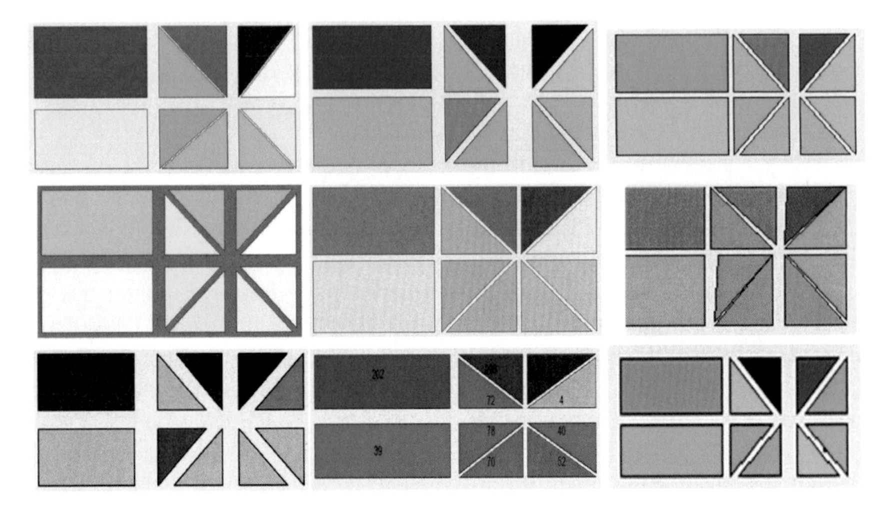

Fig. 6.4 A representation of a range of outputs from a given class of students

6.7 Expected Output from Student Work

Figure 6.4 shows example output, but with different data. It provides a range of student outputs that might be expected in a given semester. I use some class time to make an anonymized composite like this one and discuss the strengths and weaknesses of each, which gives a great opportunity to reinforce basic visual communications strategies and their effectiveness.

6.8 Assignment Variations

6.8.1 For Novice Students (Taking an Introductory Programming Course)

As I usually offer this as a second homework assignment for an introductory programming course, around week 3 or 4, I still do not expect students to know programming constructs beyond basic variable creation and use, assignment statements, and instantiating and using objects from Dr. Zelle's graphics.py library. Therefore, the novice version of this assignment is as written above.

The novice students will be able to do sorting and dragging over the range of lots in the data file to determine the total aggregate number of burials in that range of lots, so they should not need the worksheet (HW2Worksheet.docx). However, it is provided, should you wish to use it as an "unplugged" task.

6.8.2 For Intermediate Students (Taking a Python Programming Course)

For intermediate programming students, I would assign them to use a loop to input each year's data and update the fill color for each year. Each year's visualization should be a progression from the previous year's visualization (i.e., the aggregated data up to that year, not a visual representation of the individual year itself).

Here is how I introduce this more advanced assignment to my non-novice students: Your assignment is to analyze the data as above and use the results of your analysis to build a visualization of how the cemetery burials were placed over time, based on the lots in which the bodies were buried during each year. The user should click on the window to advance the image to the next year, and each year should be labelled with text inside the graphics window. The results should be cumulative, meaning that burials of previous years should still be visualized in the years following, showing the state of burials in any section of the cemetery cumulative to that year.

A handy worksheet (HW2Worksheet.docx) is provided to help intermediate and advanced students with the initial data analysis phase.

6.8.3 For Advanced Students (Taking a Software Engineering or Capstone Course)

For advanced students, in addition to the requirements mentioned in the Intermediate student assignment, above, I would assign students to use an expression to determine the change in the RGB values based on the number of variables in the lot—but using a threshold to visually allow subtle differences to be represented with the same shades, but significant differences to be represented with visually distinctive shades. For example, sectors with burials in the 60 s should not be just a slight shade different than burials in the 30 s, so a straight linear percentage would be the wrong way to go about displaying the visual difference between the two ranges.

6.8.4 For Secondary School Students

For secondary school students, I would have them use paper and progressively thinned paint (from sectors with the most burials to those with the least) to represent the density of the burials in the map sections, almost like a storyboard. After comparing their work—and making adjustments using your feedback, they can do the coding to match the drawings. Then I would assign the novice step to these students.

6.8.5 For Digital Humanities Students

Digital Humanities students could benefit from a more detailed explanation of visualization, and its role in DH, which is not within the scope of this book. They could also be motivated to explore how these data could be visualized in other ways besides the color-coded cemetery map.

Chapter 7
Textual Analysis: Frequencies and Stop Words in Dirty Text

Abstract The third of the six Digital Humanities assignments, Textual Analysis, is presented in this chapter, beginning with a brief introduction to textual analysis as an analytical method of great interest to humanities programmers. The assignment description (written in a form and with a point-of-view suitable to be copied and pasted into materials given directly to students), required support files and resources, skills utilized in the assignment (which include file I/O, string methods and indexing, and an introduction to algorithm design), assignment management techniques and issues (mostly involved with helping students learn to build mental models to deal with abstract structures, like parallel lists), atomic code for the assignment (a solar system planet exercise), expected output from student submitted work in this assignment, and variations for a number of student skill levels are provided.

Keywords Introductory programming · Programming assignment · Python assignment · Textual analysis · Literary novels · Solar-system distances

7.1 Writers Make Decisions

As I sit writing this chapter in my hammock on my back porch, on a balmy summer morning, I am conscious of a number of things: The purpose of this book (presenting six digital humanities (DH) assignments for introductory computer science students), the general approach to the topic (provide and explain a specific assignment that I currently give to my students) and, of course, the topic itself (frequency analysis of textual data). But at the edge of my consciousness are other contextual realities: the base language (American English), the stylistic assumptions that come from the conventions of the field (CS and DH educational methods), and the audience (you).

Electronic supplementary material The online version of this chapter (https://doi.org/10.1007/978-3-319-99115-3_7) contains supplementary material, which is available to authorized users.

B. Kokensparger, *Guide to Programming for the Digital Humanities*, SpringerBriefs in Computer Science, https://doi.org/10.1007/978-3-319-99115-3_7

All of these things are on my mind, and all affect my behavior as I work at the task in front of me.

Balancing all of these forces, I write one word, then another, followed by another, etc., stringing them along like popcorn garlands, in an attempt to produce a manuscript that, when published, will make sense to you and help you to successfully adopt these assignments in your own classroom (or, if you have no classroom in which to teach, to try out for yourself).

The manuscript I will eventually produce—the one you are reading now—has certain measurable characteristics. It can be measured by the length of the overall manuscript, by the number of chapters, the number of paragraphs, the average number of sentences in a paragraph, and the average number of words in each sentence.

It can also be measured by analyzing the words that I chose to get my meaning across, counting the frequencies of those words themselves. That's what this assignment is about: frequency analysis of textual data.

In this DH assignment, we focus on giving our introductory programming students the directions and resources to write a Python script to count the frequencies of individual words in a text, and to identify and report on the words that are most likely to be common among all texts written in that era.

Textual analysis in DH is a huge field that encompasses a large number of methods and approaches. This makes sense, because much of the humanities material with which DH researchers work is text-based. Though it is not within the scope of this text to provide a complete introduction and overview of digital textual analysis, the Folgerpedia site provides a list of tools, and therefore, digital-analytical approaches to texts, that are currently available to DH researchers [1]. These include concordance (focusing on the meanings of the specific words in the text), visualization (with which you are already familiar from the previous chapter), topic modelling (eliciting and analyzing topics presented by the text), frequency and statistical analysis (what we are doing here, though in a simple way), natural language processing (a branch of applied linguistics), n-gram analysis (looking for interesting word combinations), and collocations (looking for words frequently found near other words), just to name a few.

The aspect of textual analysis we are bringing to this assignment is counting the frequency of words in a text, that is, how many times a specific word occurs in that text. Some specific words stand out in certain texts, and in higher frequencies within the corpora of specific authors. But these words are often obscured by common words that occur with high frequency naturally in all texts, such as the words "the" and "and" and "he".

This DH assignment gives novice programming students the experience of downloading a text file from the web and performing word frequency analysis on it, developing an algorithm to determine "stop words" in it, and then outputting the stop words to a text file. An intermediate application of the assignment then requires the student to develop another Python program to input the original text file and the stop word file, and to then produce a list of the most frequent non-stop-words contained in the text.

Table 7.1 Word frequencies from the example sentence

Word	Frequency
I	1
will	2
not	2
my	3
son	2
who	1
is	1
really	1
land	1

7.2 The Frequencies and Stop Words Assignment

Frequency analysis in texts can range from counting total words, total number of paragraphs, and the total number of sentences. It can also consist of counting the number of times that a specific word, like the word "house", exists in a text. That is what you will be doing in this assignment.

If we have a little bit of text, like the sentence: "The quick brown fox jumped over the lazy dog," it is easy to see that each individual word in the text is presented just one time, except for the word "the", which is presented twice. Capitalization does not make "The" and "the" different words. So in a badly written sentence, such as "I will not will my son who is not really my son my land," we might get a word frequency list that looks something like that shown in Table 7.1.

So, examining Table 7.1, we see that the data that we are producing with frequency analysis is a word itself (like "will") and the frequency of that word in the text we are analyzing (a number, in this case, 2, because the word "will" is written in the sentence 2 times).

In this matter, with the help of a computer and a well-written Python program, we can count the frequencies of all of the individual words in a paragraph, a chapter, an entire text (like *Moby Dick*), or even a corpus of texts (like all of Herman Melville's collected works).

Your *Frequencies and Stop Words* assignment is to generate an "ignore list" text file (called *ignore.txt*) by opening any public-domain novel-length text, inputting the entire text into your Python program, counting the frequencies of all of the unique words in the text, and writing out to a text file the words which should be considered "stop words," or those words which should be ignored when looking at the word frequency in a text.

If you were able to process a corpus (more than one text in a collection of chosen texts), then the words of highest frequency over the entire corpus would become your list of stop words. However, since you are analyzing only one text, you might wish to use a combination of a high frequency for that specific word and word size—stop words tend to be shorter (perhaps less than or equal to 4 characters). This word size value of 4, for example, is referred to as a threshold. If you create a variable to hold the threshold, you can then change it whenever you like to “tune” it to your text.

Since you have not had the opportunity yet to become an expert in list structures, this function is given to you:

```
def inspectWord(theWord,wList,fList):
  tempWord = theWord.rstrip(“\”\’.,‘;:-!”)
  tempWord = tempWord.lstrip(“\”\’.,‘;:-!”)
  tempWord = tempWord.lower()
  if tempWord in wList:
    tIndex = wList.index(tempWord)
    fList[tIndex] +=1
  else:
    wList.append(tempWord)
    fList.append(1)
```

There are some items to note regarding this assignment.

First, this assignment (and the provided function above) uses two lists, a word list (wList) and a frequency list (fList), as parallel lists, meaning that the index references two aspects of the same data point. For example, the word “fight” might be at wList[8] and the associated frequency for the word “fight” would then be at fList[8]. An illustration to help you develop a mental picture of parallel lists based on the example sentence is provided in Fig. 7.1. You must create two list objects (wList and fList are okay) in your main() function and initialize them each to []. Then providing them (along with the specific word) as parameters in a function call to inspectWord inside a loop, that will then populate the parallel lists “automagically.”

Second, you may use any public domain text that you like. Public domain copies of novels are available for free at a number of sites on the web. You must make sure, however, that the books are in text file format. If they are formatted with “rich text” or Microsoft Word docx files, or as PDF files, or in an ePub format (or any other

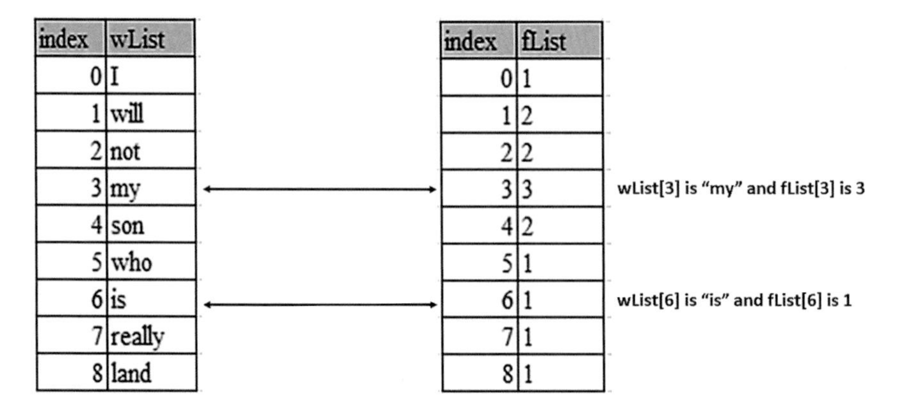

index	wList
0	I
1	will
2	not
3	my
4	son
5	who
6	is
7	really
8	land

index	fList
0	1
1	2
2	2
3	3
4	2
5	1
6	1
7	1
8	1

Fig. 7.1 Illustration of relationship between parallel lists wList and fList

format for eReaders), you will not be able to import them in a usable way into Python. Project Gutenberg (https://www.gutenberg.org/) has a "Plain Text UTF-8" download option that has worked well for students in the past. You may wish to strip out some of the front and back material that is not part of the narrative itself. Depending upon the amount of supplemental material provided by third parties, this front and back matter may skew the frequencies considerably and give false results.

Third, and directly related to the previous paragraph, is an encoding issue that may crop up when you try to read in your file to your Python program. Often your files are encoded in a different set than your version of Python is set to handle by default. The way to handle this is to set the encoding during the file open statement, like:

```
infile = open("MobyDick.txt","r",encoding="UTF-8")
```

This often suffices to take care of the problem. There are occasionally characters that, even with the specific encoding statement, are not readable and may bring your program execution to a halt. In this case, output what is read in line by line until your program stops. You will then be able to identify the line immediately before the "offending" line, and can look at that offending line in your text editor and change any strange characters to a normal character, like a normal apostrophe, double-quote, etc. Or if the offending character is punctuation, you could just delete it altogether, since punctuation is not important in this assignment.

Finally, while you are developing your program, it is important to be able to work quickly and deliberately. This is a good opportunity to examine your workflow—how you progress along with the work you need to do to solve the problem. You are advised, during your program development, to create a smaller excerpt file of your chosen novel, of around 500 words. Then as you run your program multiple times during development, you are not waiting around for 25,000 words to be processed when you discover that you made a little semantic error in your code. Since the expected frequencies of your stop words are going to be grossly different between a 500-word text and a 25,000-word text, you may need to adjust your frequency threshold in your program when you move from the development step to the "testing under load" step (that is, testing with the whole novel, not just the excerpt of the novel).

7.3 Assignment Files and External Resources

Starter Code File (if needed)

7.4 Skills Utilized in This Assignment

This assignment is designed to help introductory programming students to work intensively with file I/O and expressions inside loops. It also introduces the students to the concept of parallel lists, and list indexing.

New Programming Skills Utilized:

- File input from a public domain web source
- Cleaning dirty text (unrecognized characters, etc.)
- Using string methods to trim, replace values
- Creating an algorithm to recognize stop words in text
- Using indexing to access parallel list values
- File output using the print(<str>, file = outfile) method
- Intermediate: Managing file input from two sources

New Digital Humanities Skills Utilized:

- Exposure to real-life public domain texts and their web sources
- Dirty text and the issues of cleaning
- Frequency analysis of textual data
- Concept of stop words and strategies to identify and ignore them

7.5 Assignment Management Techniques and Issues

This assignment should be given to your students nearly halfway through the semester, when they have some familiarity with decision structures (though they would not have to have mastery over that topic yet). Here are some tips for introducing this assignment and setting students up for success.

This assignment uses parallel lists, so a considerable amount of whiteboard talk should be utilized to give the students a mental model of the structure. Lists (or arrays) are among the most difficult concepts for introductory programming students to grasp. Even your best students may have difficulty navigating the parallel list structure, although the atomic code exercise should give them enough direct experience to successfully complete the assignment. One good thing about this assignment is that students really do not need to build the parallel list structure—the given code builds the lists. All the students are required to do (famous last words, I know) is to call inspectWord in a loop, and then to traverse the newly-built lists with a decision structure (if-statement) that is true if both the frequency of a given word is greater than the set threshold and the word length of that word is less than its respective threshold.

As mentioned in the assignment narrative above, even though students should be encouraged to use texts which they like, they should be instructed to use much shorter versions of those texts (with maybe just 500 words), during the development phase to simplify development and to cut development time down. This assignment is a great opportunity to discuss project workflow, and develop good habits employed by all good programmers, such as frequently saving text, backing up files, testing small increments in code before they become big debugging problems, and “verbose mode” (that is, writing in plenty of print statements giving variable contents and process statements so that the data and progress can be checked in the case of errors—especially semantic errors, which are often difficult to track).

For intermediate and advanced students, who may then wish to use the ignore words text file to do further frequency analysis on their texts (see below), they sometimes try to solve both the problem of identifying the stop (ignore) words and the problem of doing frequency analysis of a text while ignoring the stop words in the same program. I discourage this practice, because it unnecessarily complicates the program for both the student and for you (trying to grade these is much more difficult). Plus, these kinds of two-part steps—processing to generate a text file, and then using the text file in a second step for additional processing—are common in DH data processing and a common approach to complex problems. Teaching the students to do “first things first” and to conform to requirements as written is an important part of what we do, I believe, as introductory programming instructors.

The students will also inevitably complain about “dirty” texts, or those scanned texts in the public domain which consist of inaccuracies and typos produced by optical character recognition (OCR) scanning engines (which are not perfect) and human error. There are a number of ways to deal with dirty text, but in general, the student can either leave the OCR errors in (since frequencies on a 25,000-word document will not be altered appreciably by a handful of misinterpreted letters), or correct them (a laborious task, but worthwhile to get a more precise result). The result of the assignment will not differ much either way. As mentioned above, individual characters and symbols that wreak havoc in the processing phase may need to be manually changed in the text file, or caught and dealt with (usually dropped on the floor) as exceptions.

Finally, I usually grade student submissions on this assignment by using a text with which I am familiar. For the same familiar text, I know the general frequencies of the stop words, and the frequencies of other words, especially those high-frequency words that are unique to that specific text. All that you need to do in grading is change the file name that is read in to reflect your familiar text, save it, and run it with your known text file in the same directory.

7.6 Atomic Code for This Assignment

New Planet in the Solar System

Table 7.2 shows the average distances of the planets in our Solar System from the sun, taken from the Universe Today website [2].

Suppose that you discovered a new planet in our solar system, and therefore had naming rights to it. The number of astronomical units of your newly-discovered planet is equal to your birthday. Therefore, if your name is Jenny, and your birthday

Table 7.2 Distances of planets from the sun in Earth's solar system [2]

Planet	Miles	AU[a]
Mercury	35 million miles	0.387
Venus	67 million miles	0.722
Earth	93 million miles	1.00
Mars	142 million miles	1.52
Jupiter	484 million miles	5.20
Saturn	889 million miles	9.58
Uranus	1.79 billion miles	19.2
Neptune	2.8 billion miles	30.1

[a]*Note* AU means "Astronomical Unit," the average distance from the Earth to the Sun

```
Python 3.4.2 (v3.4.2:ab2c023a9432, Oct  6 2014, 22:15:05) [MSC v.1600 32 bit (In
tel)] on win32
Type "copyright", "credits" or "license()" for more information.
>>> ================================ RESTART ================================
>>>
What is the name of your new planet? -> Brian
Please enter the month of your birth (e.g., 2 for Feb, 10 for Oct)-> 11
Please enter the day (as a number) of your birth -> 20
Your new planet is 11.2 Atomic Units (AUs) from the sun.
Your new planet is between Saturn at 9.58 AUs
and Uranus at 19.2 AUs
>>> |
```

Fig. 7.2 Input and output example for the new planet problem

is on October 19th, the astronomical units from the sun for the planet named Jenny would be 10.19, placing the planet called Jenny in our solar system between Saturn and Uranus. Write a program that utilizes these pre-set parallel lists:

planetNames = ["Mercury","Venus","Earth","Mars","Jupiter","Saturn","Uranus","Neptune"]
planetAUs = [0.387,0.722,1.00,1.52,5.20,9.58,19.2,30.1]

and searches through the AU values of the planets to determine the placement of the new planet in our solar system. The running of the program should look like Fig. 7.2.

If you need some help with the decision structures to find the proper placement of your new planet in the parallel lists, you may use this function:

```
def findPlacement(newAU, planetNames, planetAUs):
  print("Your new planet is",newAU,"Atomic Units (AUs) from the sun.")
  for i in range(len(planetAUs)):
    if newAU<planetAUs[i]:
      print("Your new planet is between",planetNames[i-1],"at",planetAUs[i-1],"AUs")
      print("and",planetNames[i],"at",planetAUs[i],"AUs")
      return
```

7.7 Expected Output from Student Work

Figure 7.3 reveals an example of a portion of the text file produced by student submissions when run with Edgar Allen Poe's *The Cask of Amontillado*. Note that the word and frequency for each word are written to the file. Some students may choose to only write out the words themselves, which would make it easier for intermediate or advanced students to input and use the file for further processing.

Fig. 7.3 Portion of the ignore.txt file produced by analyzing Poe's "The Cask of Amontillado"

7.8 Assignment Variations

7.8.1 For Novice Students (Taking an Introductory Programming Course)

The novice version of this program involves generating the "ignore" file, which requires running a loop, file input, file output, creating and updating variables, as well as utilizing built-in string and given functions. Though this will be a challenging program for novices at this stage of their programming careers, it is a good opportunity for letting them sink their teeth into something real, and to deal with a large amount of data in an elegant way, as well as a good opportunity to take pride in viewing discernible results.

7.8.2 For Intermediate Students (Taking a Python Programming Course)

The problem for intermediate students is an addition to the basic "ignore" file list described above, except that the ignore list that was generated should then be used (in a separate Python file) to eliminate words that should not be considered as high frequency words. The students must make a few changes to the given inspectWord function in their second Python module in order for it to utilize the ignoreList to choose words to ignore. The output would then be the words with the most frequency in that text which are not in the ignoreList.

7.8.3 For Advanced Students (Taking a Software Engineering or Capstone Course)

Advanced students should be able to take the requirements in the novice and intermediate steps above and apply them to a corpus of texts—a collection of related text files. In this scenario, the algorithm for discovering stop words changes from word frequency (and perhaps word size) in 1 text to a comparative word frequency in a number of texts.

7.8.4 For Secondary School Students

These methods may be used on any text—including young adult literature. As an instructor, you may get more excitement and buy-in by scanning and saving as text files one or more of your students' favorite books, or those currently on the best-seller list of young adult (YA) books. Please be attentive to any applicable copyright limitations in terms of distribution or providing public access to scanned files. Even the best scanning and OCR engine will provide some "dirty text" with which your students must deal, and the opportunity to think about word frequency in a book in which your students are highly familiar may help them make some striking connections and produce significant insights into the construction of their best-loved books.

7.8.5 For Digital Humanities Students

Digital Humanities students would be motivated to dig more into the questions of what is meant by a "text" and by a "word." Is the word "hasn't" the same as the word phrase "hast not"? In addition to simple frequency analysis, there are a number of freely available tools on the internet, which are helpful for textual analysis. In their future careers, DH students will most likely not have to generate their own tools out of code. So what is the value of having these students complete this assignment? My answer is, to help them become familiar with the textual processing methods that all tools must use to accomplish their tasks. And upon being familiar with the methods, students will then be familiar with the issues of textual analysis, and the choices made by the tool builders to address those issues.

References

1. Digital tools for textual analysis. Folgerpedia, Washington, D.C. https://folgerpedia.folger.edu/Digital_tools_for_textual_analysis (2013). Accessed 20 June 2018
2. How Far Are The Planets From The Sun? Universe Today. https://www.universetoday.com/15462/how-far-are-the-planets-from-the-sun/ (2014). Accessed 20 June 2018

Chapter 8
Code Transformation: From XML to Stylized HTML

Abstract The fourth of the six Digital Humanities assignments, Code Transformation, is presented in this chapter, beginning with a brief introduction to digital editing and archiving as a computational method of great interest to humanities programmers. The assignment description (written in a form and with a point-of-view suitable to be copied and pasted into materials given directly to students), required support files and resources, skills utilized in the assignment (which include additional algorithm development, designing and implementing branching statements, and using code to write code output), assignment management techniques and issues (mostly involved with helping students learn how to use an iterative workflow to attack a difficult problem), atomic code for the assignment (a small transformation of a classic novel exercise), expected output from student submitted work in this assignment, and variations for a number of student skill levels are provided.

Keywords Introductory programming · Programming assignment
Python assignment · Code transformation · XML-to-HTML transformation
Novel personalization

8.1 The Magical Aura Around Computers

I first started programming computers in the 1970's, because I was enthralled by the mystique of these electronic devices that could do—to the untrained eye—magical things. Television shows and movies during that time depicted computers as autonomous metal boxes with blinking lights, which merely needed to be asked a question and would then "spit out" (sometimes literally) the perfect answer. This seemed like a magical process to me, and it did not take long for me to realize that

---

Electronic supplementary material The online version of this chapter (https://doi.org/10.1007/978-3-319-99115-3_8) contains supplementary material, which is available to authorized users.

B. Kokensparger, *Guide to Programming for the Digital Humanities*, SpringerBriefs in Computer Science, https://doi.org/10.1007/978-3-319-99115-3_8

I could harness this magic—conjure it, in a sense—by simply understanding what was going on inside the box and communicating in its language.

Of course, there is nothing magical about computers at all—they are simply electronic devices that store and process data. But I still see that look of awe in users' faces when I unveil an application that helps them solve a problem, and does so in an effective way.

This assignment captures that aura of magic with students as they write code to transform complex XML (Extensible Markup Language) into simple HTML (Hypertext Markup Language), stylized with a very simple CSS (Cascading Style Sheet) file. This transformation process is used (though with a mature set of tools described briefly below) by a number of publishers to push content out in a number of formats. In fact, the publisher of this monograph (Springer) has a well-developed workflow for this process [1]. This assignment gives students a gentle introduction to the rich field of digital editing and archiving.

The process of digital editing and archiving of texts has a well-designed workflow, which generally begins with encoding the document in XML using standardized protocols, such as the TEI (Textual Encoding Initiative) template. The richness of the XML markup then allows editors to use XSLT (Extensible Stylesheet Language Transformations) to transform the XML document into other formats, such as HTML, ePub, and PDF, for publishing in different media.

Though any serious DH scholars or digital archivists should immerse themselves into TEI and XSLT, I believe that it is important for any DH practitioner to understand the basic process through which transformation occurs. The experiences offered to students through this assignment do not exactly mimic the exact transformation process handled with a transformation engine like that which employs XSLT, but instead asks students to use a number of decision structures (if-statements) to identify XML markup code and, in response, to apply the correct HTML markup code in its place.

Ultimately, this is an excellent way to teach introductory students how to use decision structures and to see the effects of both their correct and incorrect choices they make when designing those decision structures. The fact that DH students encounter some of the issues and challenges of the digital editing and archiving process is a convenient bonus to their DH educations.

8.2 The Code Transformation Assignment

In the digital editing and archiving of texts, it is common to markup the text in XML (Extensible Markup Language) using a certain protocol, like TEI (Textual Encoding Initiative), and then use a transform script to translate the XML Markup tags to the needed format. When a publishing organization, such as the Folger Shakespeare Library, wants to put the text up as a web page, it uses a transform script to translate the XML tags to simple HTML tags.

```
WTExcerpt.xml
4  <head>
5  <w> ACT </w>
6  <w> 1 </w>
7  <w> Scene </w>
8  <w> 1 </w>
9  </head>
10 <stage>
11 <w> Enter </w>
12 <w> Camillo </w>
13 <w> and </w>
14 <w> Archidamus </w>
15 <pc>.</pc>
16 </stage>
17 <sp>
18 <speaker>
19 <w> ARCHIDAMUS </w>
20 </speaker>
21 <w> If </w>
22 <w> you </w>
23 <w> shall </w>
24 <w> chance </w>
25 <pc>,</pc>
26 <w> Camillo </w>
27 <pc>,</pc>
28 <w> to </w>
29 <w> visit </w>
30 <w> Bohemia </w>
31 <w> on </w>
```

Fig. 8.1 A small portion of the simplified XML file provided in WTExcerpt.XML

Your task, for this assignment, is to use Python to:

- write a module that inputs an XML file excerpt from Shakespeare's *The Winter's Tale*,
- translates the inputted text to HTML tags so that a simple CSS file (provided below) can influence the display of the content in the browser, and finally,
- writes the HTML file out to disk.

Here are some details you will need:

Figure 8.1 shows a small portion of the simplified XML code for the provided *The Winter's Tale* excerpt. A link to the file, WTExcerpt.xml, is provided in the Assignment Files and Resources section below. To view it, open it in any text editor. Note that XML (even the very simplified version of XML you see here) has some similarities with HTML, but often the tags are used differently. For example, in an HTML file, there is usually one <head> element, near the top of the document, that sets up the environment and code for the page. In an XML file encoded in TEI, there can be a number of <head> elements, as the TEI description of the <head> element is a: "(heading) contains any type of heading, for example the title of a section, or the heading of a list, glossary, manuscript description, etc."[1]

Figure 8.2 shows the simple CSS file (script) provided for this assignment. A link to the file, Shakespeare.css, is provided in the Assignment Files and Resources section below. Please note that it is a simple CSS design. The point of this assignment is not to learn how to write awesome CSS code and it is certainly not about learning webpage design. You are simply asked to use the provided CSS file to style the HTML tags that you choose to markup your content in your HTML version of *The Winter's*

[1]TEI stands for Textual Encoding Initiative. Information about TEI is available at http://www.tei-c.org/release/doc/tei-p5-doc/en/html/ref-head.html. The xml excerpt file provided for this assignment is not an accurate version of TEI-coded XML, but instead is simplified to facilitate a good learning experience for the students. A good assignment for advanced students would be to use the full and accurate TEI-coded XML file available at the Folger Digital Texts website.

```
Shakespeare.css
1  body {     font-family: arial;     }
2  h1 {     background-color:#CCC;
3    border: 1px solid;
4    color:#880000;
5    text-align: center;     }
6  h2 {  color:#FF0000;
7    text-align: center;     }
8  h3 {     background-color:#8888FF;
9    border: 1px solid;
10   color:#AA0000;
11   font-style: italic;     }
12 h4 {     color:#880000;
13   font-style: italic;     }
14 p {     color:#000000;          }
```

Fig. 8.2 The simple CSS script provided for this assignment denoting the limited number of HTML elements that the students may use

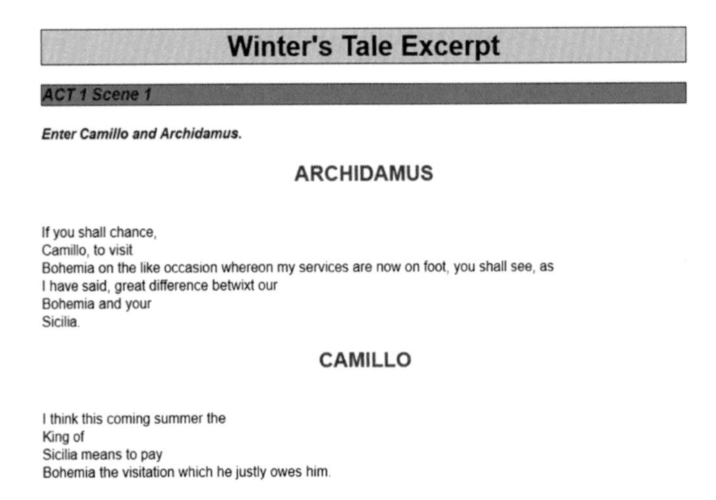

Fig. 8.3 An example of how the beginning of *The Winter's Tale* excerpt should look in a browser after being transformed to HTML by the student and stylized by the provided CSS file

Tale excerpt. You must link the CSS file in the <head> of your generated HTML file, like: <link rel = 'stylesheet' href = 'Shakespeare.css' type = 'text/css'/>. You should become familiar with the provided CSS file, because your translation of the XML file into HTML may only use the HTML tags that are styled in the CSS file.

When you have run your translation program in Python, and generated and saved the resulting HTML file, open it in a browser. It should eventually end up looking like the web page screenshot provided in Fig. 8.3.

Before you begin inputting and processing the characters (speakers) and their lines from the XML file, it is often helpful to open the output file and write to it the header information for the file (the link to the CSS file, the <title> tag, etc.). After you have processed the XML file and transformed the tags to HTML tags, don't forget to print out to the file the end of HTML file information, such as the end tags for </body> and </html>.

Though the savviest students among you might realize that you can complete this assignment using simple string.replace(string,string) function calls, the point of this assignment is for you to practice using branching (decision) statements. Therefore, the solutions that receive full credit will use if, if-else, and/or if-elif-else statements to process the transformations.

Please name your HTML output file <yourlastname>WT.html. When I view your work, I will use the XML file and CSS file as given in the assignment, so please do not change them.

When you are ready to upload your work, simply upload the Python module as <yourlastname>Transform.py by the stated deadline. There's no need to upload anything else for this assignment.

Please remember, as usual, the appropriate header information for your Python module, to use self-documenting identifiers, and to make your input and output user-friendly.

8.3 Assignment Files and External Resources

WTExcerpt.xml (adapted from The Winter's Tale [3])
Shakespeare.css
ExampleFormat.PNG
Starter Code for Novices.

8.4 Skills Utilized in This Assignment

This assignment is designed to help introductory programming students use their newly-learned logical skills in forming branching statements, and reinforces their file input and output skills. Digital Humanities students gain familiarity with XML-encoded documents (though the excerpt used in this assignment is a simplified version of an actual richly-encoded XML document).

New Programming Skills Utilized:

- Planning a strategy for transforming a file from format A to format B
- Designing a series of branching statements to implement the strategy
- Forming proper if, elif, and else statements in Python
- Using Python to output a file with simple HTML markup
- Linking an existing stylesheet to outputted HTML code
- Managing simultaneous input and output files in Python.

New Digital Humanities Skills Utilized:

- Exposure to real-life XML encoding with the TEI template
- An under-the-hood view of file transformation
- A familiarity with HTML markup
- A familiarity with linking CSS files for formatting HTML
- Exposure to the Folger Digital Texts Online resource.

8.5 Assignment Management Techniques and Issues

This assignment should be made sometime early in the second half of a semester course, after decision structures have been thoroughly introduced.

Students must be a little familiar with HTML to do well on this assignment. If a large number of students in your introductory programming class do not know HTML, you could use this as a "pairs programming" assignment and pair up students who do not know HTML with those who are more familiar with it.

Students do not need to know CSS coding—in this assignment, they are given a CSS file and the actual code needed to link and use that file. In regard to the provided link, as long as the CSS file is in the same directory as their generated HTML file, that link should work to style any HTML elements listed in the CSS script. Students are not permitted to alter the CSS file in any way. I warn them that I will use my base files (XML and CSS) in grading their assignments, so if they alter these files, their submitted Python code will likely not work well (and subsequently, they will lose points on the assignment).

Students are often tempted to upload and submit the entire project (all files) in a zipped file. This is not necessary—it only takes up more storage space. Students should be advised in written materials as well as orally that they need only upload one effective Python (.py) file to satisfy the requirements of this assignment.

Some students will avoid using branching structures by using a series of *replace*(string,string) statements. If done correctly, this approach will actually produce the transformed file, and satisfy the assignment requirements. But this approach gets around the branching structures practice that is so vital to this assignment. I make it clear to students that they will be docked points if they use replace statements to avoid using branching statements.

This assignment provides a large number of challenges for the introductory programming student, and I advise my class to take an incremental mentality when working on it. Odds are that most novice students will not come up with a perfect solution to this assignment. Many will approximate the requirements, falling short on most or all of its aspects. I advise the students to do the task/code/review approach:

- Decide, from previous code additions, what needs to be "fixed" next, for example, properly code the speaker

- Write code, and revise that new code as necessary, to fix that specific problem, for example, write an if-statement to compare the given line to the XML markup denoting the speaker, and write out the associated HTML tag in its place
- Save and run the code to see if the fix worked, for example, when refreshing the HTML file in my browser, is the speaker now displayed as required?

Then they repeat this process with their allotted programming time, until they have either produced a transformation script that satisfies all aspects of the assignment completely, or until the submission deadline is reached, whichever comes first. This allows the students with marginal programming skills to still submit a solution (with very little point loss on the assignment), and those with more developed programming skills to submit a more comprehensive solution.

8.6 Atomic Code for This Assignment

The Novel You

In this atomic code assignment, you will use the same text file of a novel that you used for the *Frequencies and Stop Words* assignment in the previous chapter. If you did not do that chapter, just choose any text file version of a novel in the public domain and download it to your computer to use. As in the previous chapter, I will use Melville's *Moby Dick* [2] as an example.[2]

For this atomic code experience, you are to input the text file of your novel, and substitute your name for the main character of the book. So wherever that character's name appears in the text, your name will now appear in its place. With that done, then revise your script to substitute the names of your family members and friends for other characters in the book as well. You could change place names in the book with place names near you, etc. In other words, have fun with this assignment!

When your Python script has completed the substitutions, you should write it out to a new text file, with the original title of the novel and "Me" at the end, followed by the.txt extension. So MobyDick.txt would be outputted, with your substitutions, as MobyDickMe.txt (see Fig. 8.4 as adapted from Moby Dick found on the Project Gutenberg website 2018).

---

[2]The text file for Moby Dick provided at Project Gutenberg at https://www.gutenberg.org/ebooks/2701, provides a great deal of front matter, which I stripped out so as to avoid skewed results for textual processing. I recommend you do the same for any file that you use for processing.

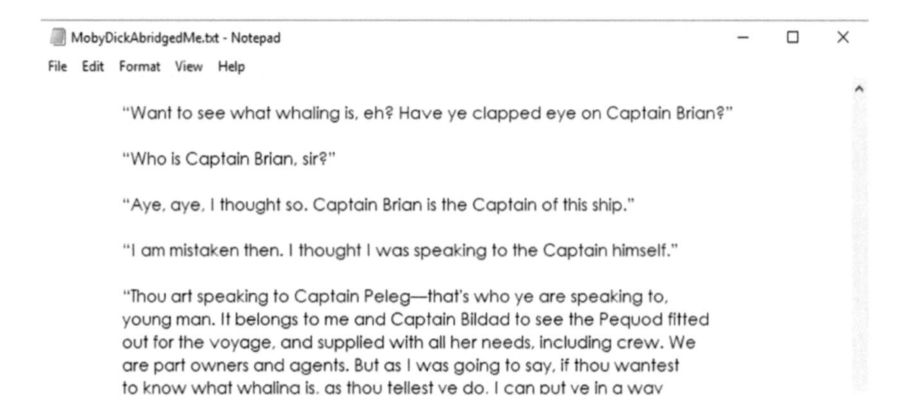

Fig. 8.4 A small excerpt from Moby Dick transformed with my name as the Captain Ahab character

8.7 Expected Output from Student Work

The expected output of the solution is provided for the students in the Example-Format.PNG file, with a link available in the course resources section above, and illustrated in Fig. 8.3.

8.8 Assignment Variations

8.8.1 For Novice Students (Taking an Introductory Programming Course)

Novice students will find this assignment challenging, most likely the most challenging among all six assignments. By this point in the semester, I often discover that they still have not developed a good mental model for the file I/O, which bites them on this assignment. Also, many novice programmers do not have any exposure to HTML, so they find themselves struggling with the syntax of three languages—Python, XML, and HTML. For that reason, I usually employ a pairs programming approach for this assignment for a classroom full of novice students, and provide some starter code that shows them how to output the <head> element, begin the <body> element (complete with the link element written out to the HTML file as provided), and write the <h1> tag around the stated title in the XML file, all outputted to the HTML output file. I also explain to them that this project should be done by degrees. Get the entire XML excerpt file printing out to the HTML file first (even though the formatting will not be correct), and then incrementally add code to their Python files to improve the formatting of their output files. When they have done it to their satisfaction, OR they have reached the assignment submission deadline (whichever comes first), they should submit the Python solution and move on to the next adventure in the course.

8.8.2 For Intermediate Students (Taking a Python Programming Course)

Intermediate students should be able to easily handle the decision structures for this assignment, so for them the challenge might simply be the HTML. For this group, the output HTML file should be perfect, including dealing with strange characters (such as exists in the title of the excerpt). Intermediate students should work independently (no pairs programming) on this assignment, though be prepared to provide a brief introduction to HTML and CSS in the classroom, for those who need it, or to point the students to web resources that will provide a good introduction.

8.8.3 For Advanced Students (Taking a Software Engineering or Capstone Course)

Advanced students should have no trouble completing this assignment as stated, even if they don't know HTML or have any familiarity with CSS. If this is the case, they should be expected to do what all big boys and girls do and find resources to teach themselves about HTML. I would not hesitate to give advanced students the simplified XML version of the assignment (as detailed above), have them complete that part as an introduction to the concept, and then give them the entire actual XML file to transform. Advanced students should be able to handle that level of complexity.

8.8.4 For Secondary School Students

Secondary school students may need lots of hand-holding for even the simplest of assignments in this area. I personally would hesitate to use this assignment in the secondary school classroom, unless in a talent and gifted program, with students who already have familiarity with HTML and CSS. If not, it is my humble opinion that the learning curve is just too great for students to succeed with this assignment. You may be better off making this assignment an opportunity for you to demonstrate your programming skills to your class.

8.8.5 For Digital Humanities Students

Digital Humanities students may find this assignment unnecessarily tedious—they will most likely be aware of applications that do these transformations for them, and may not see the point in learning to write code to treat a simplified version of the problem. If I had a classroom composed entirely of DH students, I would probably lead them through coding in TEI and basic XSLT for this assignment, in an application like Oxygen.[3] Though this does not give them Python coding experience, it does begin to give them familiarity with a useful skill, which students interested in digital editing and archiving will certainly need in the future. So that they do not miss out on decision structure practice, I would then give them Python coding experiences with more straightforward decision-structure assignments, many of which are available in any introductory programming textbook.

References

1. Manuscript preparation. Springer. https://www.springer.com/gp/authors-editors/book-authors-editors/manuscript-preparation/5636 (2018). Accessed 20 June 2018
2. Moby Dick; Or, The Whale by Herman Melville. Project Gutenberg. https://www.gutenberg.org/ebooks/2701/2701-0.txt (2018). Accessed 20 June 2018
3. The Winter's Tale. Folger Digital Texts, Washington, D.C. http://www.folgerdigitaltexts.org/download/ (2018). Accessed 20 June 2018

[3]https://www.oxygenxml.com/.

Chapter 9
Art Stylometry: Recognizing Regional Differences in Great Works of Art

Abstract The fifth of the six Digital Humanities assignments, Art Stylometry, is presented in this chapter, beginning with a brief introduction to stylometry as an analytical method of great interest to humanities programmers. The assignment description (written in a form and with a point-of-view suitable to be copied and pasted into materials given directly to students), required support files and resources, skills utilized in the assignment (which include additional object instantiation and use, additional familiarity with images, and looping structures), assignment management techniques and issues (mostly involved with helping students learn how to properly use starter code to get started and then to move through the problem with their own code), atomic code for the assignment (an image processing exercise), expected output from student submitted work in this assignment, and variations for a number of student skill levels are provided.

Keywords Introductory programming · Programming assignment
Python assignment · Stylometry · Stylometric analysis · Regional differences
Art masterpieces · Leonardo DaVinci · Rembrandt

9.1 Comparing Features in Artifacts

I am a fan of Dr. John Zelle and his textbook, *Python Programming: An Introduction to Computer Science* [1]. The inspiration for this assignment came from one of the programming exercises in the book, where Dr. Zelle provides some basic code, his graphics.py library, and a "brightness" formula, and then asks the student to traverse every pixel in an image to create its grayscale equivalent, saving the new image to disk as a separate file.

Electronic supplementary material The online version of this chapter (https://doi.org/10.1007/978-3-319-99115-3_9) contains supplementary material, which is available to authorized users.

B. Kokensparger, *Guide to Programming for the Digital Humanities*, SpringerBriefs in Computer Science, https://doi.org/10.1007/978-3-319-99115-3_9

The possibilities of what can be done using this image object, with the ability to visit and process one pixel at a time, blew my mind. "Why not use this to measure and compare color averages in great works of art?" I wondered, and the Art Stylometry assignment was born.

Most anyone who has even an introductory course in art history can recognize major periods and regional differences in great works of art, by simply looking at exemplars among them. What we are doing for this digital humanities (DH) assignment is comparing style among great works of art. Style is broadly defined by the Merriam-Webster online dictionary as "a particular manner or technique by which something is done, created, or performed" [2]. So what the viewer recognizes when identifying a specific style is a "manner or technique," such as the composition of the piece or, more germane to this assignment, the colors used in the production (mostly influenced by the mixture of background colors) of the piece. But if humans are very good at recognizing the style of a piece of art, how do they teach a computer to recognize it? The solution lies in the twin methods of measurement and comparison of the findings.

There is a rich history of stylometry, much of which is directed towards author attribution (hypothesizing as to the creator of a target work based on known features of that creator's other known works). Though it is not within the scope of this book to provide an in-depth discussion of stylometry in the DH field, there have been a number of high-profile stylometric studies, including a number provided in an excellent tutorial regarding stylometry in Python [3].

Stylometric Analysis is used to determine artist and author attribution, as well as to identify features important to specific eras and geographic places. Assignment 5 directs students to use the image methods provided in Dr. Zeele's graphics.py library[1]to analyze 9 portrait images (provided in a zipped file in the assignment materials) for specific feature values, then requires students to write Python code that makes a determination as to which artist—Leonardo Da Vinci or Rembrandt—the measured features of the three mystery portraits most resembles.[2]

9.2 The Art Stylometry Assignment

Stylometry is a research method used by Digital Humanities scholars, among others, to train the computer to recognize and identify style markers among various artifacts, such as manuscripts, texts, and—in the case of this assignment—great works of art.

[1]Dr. Zelle's excellent graphics library can be found at http://mcsp.wartburg.edu/zelle/python/, which provides a link to his graphics.py module and documentation. If you do not use this graphics library, there are other libraries you can use, or develop your on using the TKInter library (information available at https://wiki.python.org/moin/TkInter).

[2]The photographic images made as faithful reproductions of art images provided for this assignment, and used as illustrations in the discussion, are public domain images, and thus may be used freely in publications. More information regarding the rights associated with these specific images is available at: https://commons.wikimedia.org/wiki/Commons:Reuse_of_PD-Art_photographs.

One way of recognizing style among artwork is to measure the choice and usage of color in paintings. Some paintings use warmer colors (generally understood as reds and oranges), while others use cooler colors (generally understood as blues and dark greens). Although individual artists have ultimate freedom to choose any colors in any given moment while working on a painting, these artists work in a cultural context, which provides a cultural expectation of style that looms over their final products. If working for a patron, the artists must satisfy their patrons' stylistic expectations for the piece. In short, "weird" paintings that defy the style of the time are at risk of being rejected. Working for sales in galleries is even more risky, if the artist strays too much from the stylistic conventions of the time. In some cases, the artists, themselves, may have been instrumental in actually setting those stylistic conventions, so there is even more pressure to conform to them. It follows, then, that there are recognizable features of paintings that give cues to whether a painting is a Rembrandt, or a Da Vinci, since these artists worked in different regional cultures and at different eras in time. In this assignment, you will measure a number of paintings for these cultural "cues", and determine which style is most similar to those of three "mystery" paintings.

For this Art Stylometry assignment, you will create two modules:

One module will "train" your system on the color choice differences between the artistic works presented by two master artists: Rembrandt Harmenszoon van Rijn (Rembrandt) and Leonardo di ser Piero da Vinci (da Vinci).

A second module will use the training data provided by your first module to compare the averages for the color choices in the training data with the color choices provided in a "mystery" painting. Once the comparison is made, the second module will then identify the training work that is most similar in color makeup to the "mystery" painting.

Module 1—Training Module

Create a module that loads each of the six images (the three DaVinci paintings and three Rembrandt paintings) provided in the zip file of images (link available in the Assignment Files and External Resources section below) and uses a loop to accumulate the total red, green, and blue (RGB) values of each pixel into corresponding accumulator variables. After running the loop, the average values (total accumulated amounts/number of pixels) for red, green, and blue, should be stored in a text file for each of the six paintings. You may use the tools available in Dr. Zelle's graphics.py library or build your own using TKInter.[3]

Your training module should perform these tasks:

- For each painting, traverse the pixels, analyzing each image for these three features: Average value computed from each of the sum total values for red, green, and blue over all pixels
- Store the filename and averages for each of the six paintings in a text file for use in module 2. The file should look like the example in Fig. 9.1, which shows the data format but with made-up data (not actually what your module will produce).

[3]Zelle, op. cit

artFeatures.txt - Notepad

File Edit Format View Help

```
DaVinci1.gif        70.941005         55.8629   38.350135
DaVinci2.gif        47.667705         40.50896  15.96666
DaVinci3.gif        77.92661  61.694945         40.112325
Rembrandt1.gif      57.363675         42.440505           23.43765
Rembrandt2.gif      71.462865         54.773835           22.47438
Rembrandt3.gif      86.603295         52.546255           13.14048
```

Fig. 9.1 Example output file for the training module with made-up data

Module 2—Speculation Module

After the training module has been completed, and a suitable output file generated (like in Fig. 9.1), now create a module that does these tasks:

- Load in the data from the training module file generated (example in Fig. 9.1).
- Load the mystery portrait and analyze it in the same manner that you did for the training images in Module 1, above.
- Compare each of the features in the mystery portrait (average red, average green, and average blue) with those average values of the six training portraits, using a simple distance formula on each dimension:
  distance = | trained value−mystery value | /mystery value.
- Determine which is the "closest" portrait to the mystery portrait, by adding all of the computed distances together and determining the smallest sum from among all six.
- Display both the mystery portrait and the closest training portrait and output the associated values for both.

When you have completed both modules, zip ONLY the two Python modules themselves (not the training or mystery images, not the graphics.py file, and not even your training data output file) and submit them by the deadline. When I grade your files, I will ensure that all of the training and mystery images are in the same directory as your modules, and will run your module 1 to generate your training data, and then immediately afterwards, I will run your module 2 on one of the mystery images to view your output, which should show me the training image which is most similar in color choices to my chosen mystery image. I may use one of the mystery images that was provided to you for your testing, or I may use an entirely different mystery image in the grading process.

9.3 Assignment Files and External Resources

Assignment 5 Images ZIP File
HW5 Starter File (if needed)
Extended Starter File (if needed).

9.4 Skills Utilized in This Assignment

This assignment is designed to give introductory programming students experience in forming and controlling loops, as well as to reinforce all of the previous structures learned.

For DH students, this assignment is designed to give DH students an introduction to stylometry, though with world-famous artworks as an alternative to the way it is normally taught (i.e., with text files). This broadens DH students' scopes beyond textual data, and also gives them a general introduction to programmatically working with images.

New Programming Skills Utilized:

- Reinforcement of using objects but also at a higher level
- Additional familiarity with working with graphics files in Python
- Creating and controlling loops in the Python language
- Additional file input and output experience
- Modifying starter code to solve a problem.

New Digital Humanities Skills Utilized:

- Exposure to stylometry as a DH research method
- Familiarity with how images are digitally stored
- Practice generating measurements on image data
- Familiarity with the concept of training modules and speculative modules
- Exposure to some of the measurable dimensions of classic art pieces.

9.5 Assignment Management Techniques and Issues

By this time in the typical introductory programming course, students should have been introduced to all of the primary structures of the Python programming language, including how to construct and control loops. So far, we have taken it easy on the novice students, by requiring them to do initial steps of the previous assignments, reserving the full implementation of the assignment tasks to intermediate or advanced students. The "kid gloves treatment" ends with this assignment. Even novice students are expected to do a full implementation of this assignment, which will test their mettle both in programming technique and coding skills as well as in algorithm design to solve a problem. There will, no doubt, be some "wailing and gnashing of teeth" in the introductory programming classroom during this assignment. However, no introductory programming course is complete without requiring students to take a non-trivial problem of moderate complexity (such as this one), and complete it to a satisfactory outcome (an outcome that is easy to see if and when it is done correctly).

Here are some tips for introducing this assignment and setting students up for success:

Previous to attempting this assignment, students should have already completed 1 or more exercises or assignments using the image object of the imported graphics.py module, or your chosen library to support graphics manipulation. Assignment 2, *Visualizing Change Over Time*, provides adequate experience with the graphics environment of Python to approach this assignment, but some starter code may be required if students have not used the Image object in the graphics.py library, or manipulated images in some way with Python code.

Since this is a complex, multi-step assignment, instructors should caution students to work through this assignment more out of a top-down approach, breaking the problem into a number of smaller pieces and developing from there. Some students will be overwhelmed by this assignment on outset. They may complain that they do not know where to start. The atomic code provided for this assignment gives them a good place to start. However, with this atomic code experience, you may encounter a second problem: Students may stay too close to the atomic code, and not be willing to adapt, add, and omit code as necessary to solve this specific problem. I cannot tell you the number of times that I gave my students starter code (that traverses the image to produce a red filtered copy of the image and saves it to disk), only to have the red filtered image and saving mechanism survive through to the bitter end of the assignment. This is a good occasion to discuss how to re-use code thoughtfully and legally both in the classroom and beyond its borders.

As in the previous assignments, where intermediate students and above were given multi-step tasks for a single assignment, many students will inevitably attempt to write all of the code for both modules as one big Python module. This quickly overwhelms them, especially if they attempt to merely add speculation module functionality to training module functions, and also makes debugging (and grading) difficult—if not impossible—for erroneous code. One solution for this is to have them turn in each module at different stages with different deadlines. The training module may be due, and graded with feedback, and then a little later, the speculation module may then be due. Each should produce only the functionality it is required to produce.

Beyond making this process easier to grade and debug, having students stage their submissions for this assignment may also give them the opportunity to learn the value of workflow in creating a training module and then, afterwards, an independent speculation module. As some DH students, in particular, may be working as graduate students in the field at a later date, they may only be working on the training aspect of stylometry, as a process in itself.

Fig. 9.2 DaVinci image with the red filter applied and both images displayed

9.6 Atomic Code for This Assignment

For this atomic code assignment, load a gif file into your Python module, and use a nested loop to traverse each pixel in the image. Create a new image, and for each pixel, max the red value out to 255 and apply that pixel to the new image. Display the new red-filtered image and give the user the opportunity to save the file as a new .gif image. Fig. 9.2 shows a DaVinci image with the red filter applied, with both separate images displayed side-by-side.

9.7 Expected Output from Student Work

Figure 9.1 shows an example of what the output file might look like (with made-up data for this example). Figure 9.3 shows the images that are output—both the mystery image, and its closest match from among the training images. Figure 9.4 shows the text output with the averages of all of the training images as well as the mystery image.

Fig. 9.3 Mystery and Output images displayed after running the speculation module

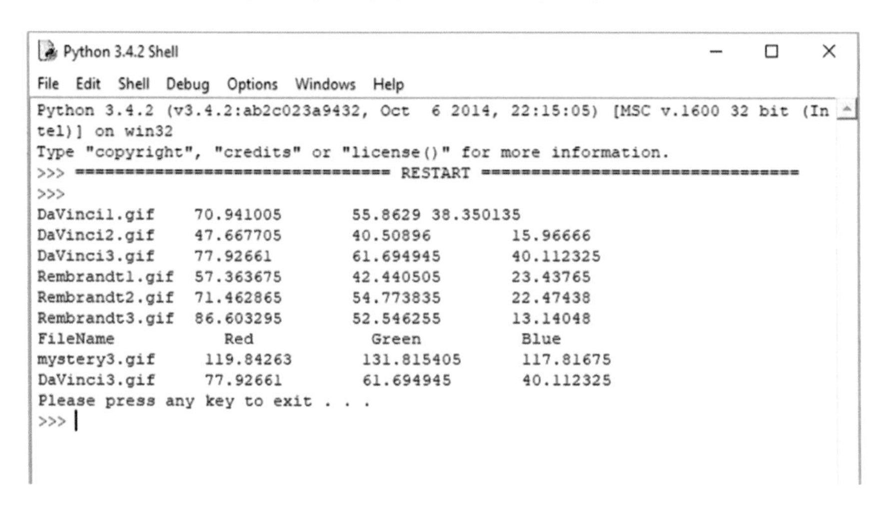

Fig. 9.4 Text output in the IDLE shell when the speculator module is run

9.8 Assignment Variations

9.8.1 For Novice Students (Taking an Introductory Programming Course)

By this point (about 2/3 to 3/4 of the way through the semester) students should be able to do a full implementation (both the training module and the speculation module) of the above assignment, though the loops handling large amounts of data and the file I/O will challenge them. We've shielded the novice students so far in the previous assignments by only requiring them to do the first steps of the full implementations, reserving the more complex later steps for intermediate and advanced students. At this point of the semester, it is appropriate to require novice students to complete the entire implementation of this assignment, which will challenge them. Once they have completed and submitted the assignment, however, novice students often feel much more prepared to attack more challenging programming problems in the future.

9.8.2 For Intermediate Students (Taking a Python Programming Course)

For intermediate students, challenge them to also analyze the paintings by evaluating the average red, green, and blue values for a rectangle that encompasses the face of the portrait (after the user clicks the upper-left and lower-right corners of the face). All other aspects (including distance measurements) should be the same. This effectively doubles the number of dimensions collected, measured, and compared (the given avgRedEntire, avgGreenEntire, and avgBlueEntire, plus avgRedFace, avgGreenFace, and avgBlueFace). Do the students' speculation modules provide different results than those that only measure the color mixes in the entire paintings?

9.8.3 For Advanced Students (Taking a Software Engineering or Capstone Course)

Advanced students should be able to create code to do this problem from beginning to end. Therefore, you should not give the advanced students the starter code. I advise you to give them the assignment, and let them design their solutions and implement them from end to end. You may also want them to choose their own dimensions for measuring features of the artwork, which may end up being different than those suggested above. Can a student come up with a good way to characterize brush strokes? You would probably need to provide an image of a higher resolution for that challenge. Or, can a student come up with a way to quantify the artist's composition of the figure in the painting, or in other works which are not portraits? Also, advanced students are not bound to Dr. Zelle's graphics.py library, but can most likely work with the native Python libraries for image manipulation. See the footnote 1 for more information.

9.8.4 *For Secondary School Students*

This will be a most challenging project for secondary school students. These students might best learn about art stylometry by estimating RGB values when projecting the images in front of the class, and then comparing their estimations to your average values when running your proof of the training software. Again, they might predict which training image will end up being most like a given mystery image, and then you can show them the speculation data and flash the two images back and forth. You are the best judge as to whether your students can handle this assignment, but even as an in-class demo, they will still learn a great deal about stylometry, and will hopefully be motivated to learn more and try out some of the techniques outside of class.

9.8.5 *For Digital Humanities Students*

Since stylometry is bread-and-butter for some DH scholars, DH students should spend as much time as possible investigating the paintings in various dimensions, and doing a little exploration in textual stylometry as well. However, as a programming course, the modeling and comparative aspects of this assignment should be emphasized as a general way to do stylometry, independent of its domain. DH students will, I am convinced, be intrigued with this assignment and will want to do more stylometric exploration on their own outside of class.

References

1. Zelle, J.: Python Programming: An Introduction to Computer Science, 3rd edn. Franklin, Beadle, Portland, Oregon (2016)
2. Style: Definition. The Merriam-Webster Online Dictionary. https://www.merriam-webster.com/dictionary/style (2018). Accessed 20 June 2018
3. Laramee, F.D.: Introduction to Stylometry with Python. The Programming Historian. https://programminghistorian.org/en/lessons/introduction-to-stylometry-with-python (2018). Accessed 20 June 2018

Chapter 10
Social Network Analysis: Historic Circles of Friends and Acquaintances

Abstract The sixth, and final, of the Digital Humanities assignments, Social Network Analysis, is presented in this chapter, beginning with a brief introduction to network analysis as an analytical method of great interest to humanities programmers. The assignment description (written in a form and with a point-of-view suitable to be copied and pasted into materials given directly to students), required support files and resources, skills utilized in the assignment (which include library imports in Python, object instantiation and use, and dictionaries as data structures), assignment management techniques and issues (mostly involved with helping students learn about the concept of code re-use, and how to install and review documenation for Python libraries), atomic code for the assignment (the in-class tutorial), expected output from student submitted work in this assignment, and variations for a number of student skill levels are provided.

Keywords Introductory programming · Programming assignment
Python assignment · Network analysis · Historic relationships · Friend circles
Networkx library

10.1 Code Reuse and Python Libraries

It is most likely true that, for any given programming problem, there already exists a solution to that problem that has been designed, implemented, and tested. Within this backdrop, the solution to a problem then becomes more focused on the attainment of written code, and less focused on writing new code to solve the problem.

I use code that has already been written by someone else (referred to as *code reuse*) in most of my programming projects. If I have a segment of code for which I do not have an immediate solution in my head, I do a quick web search and find dozens of

Electronic supplementary material The online version of this chapter (https://doi.org/10.1007/978-3-319-99115-3_10) contains supplementary material, which is available to authorized users.

B. Kokensparger, *Guide to Programming for the Digital Humanities*, SpringerBriefs in Computer Science, https://doi.org/10.1007/978-3-319-99115-3_10

example code segments from which to copy and modify for my own use. This is not wrong—most of the programming resources on the web are a large ever-changing pool of "share and share alike" code snippets that are there to help programmers solve problems. Professional programmers use these resources extensively to save time and to benefit from the knowledge, feedback, and testing results that these resources provide. When there is a particularly masterly treatment of a problem (one that I could not likely have created myself, given an appropriate amount of time and effort), I give attribution to the maker in a comment within the code itself. I do appreciate the time and effort expended by those who share their code, as I share my own code on occasion in these resources as well.

It is important, though, for introductory programming students to learn how to write code to solve problems, not to learn how to find and borrow code. These students must learn how to use all of the programming structures in Python, including loops, decision structures, lists, file input and output, and object instantiation and use. Students who jump too quickly to use code provided for them, and who do not take the time to learn how to create and develop those structures for themselves, are cheated out of their learning opportunities in the course. If they do not learn how to make these structures and use this syntax in an introductory programming course, it is doubtful that they will learn to use them anywhere else, and will find themselves underprepared for later courses. They must be held to a standard where they demonstrate the knowledge to create all course learning targets in the Python syntax in your course.

However, code reuse is such an important practice in the field that I cannot help but introduce my students to it—with the hope that they will learn to do it correctly, and with proper attribution where appropriate. I model this in the classroom when students ask questions for which I do not have an immediate answer. If the answer is also not in the textbook, I show them how to search the web, find an answer from a reputable website, copy it and paste it into a Python module, modify and attribute it appropriately, and then test it out.

It is with this spirit that I introduce this final assignment, social network analysis. Network analysis is a method of analyzing and visualizing connections between a number of entities, be they human, biological, technological, or organizational, among others. If connections can be observed among entities, they can be captured and analyzed by using a rich set of analysis and visualization methods, most of which are provided to the researcher free of charge, with existing libraries. Since these tools are ever-changing—some arise daily, while others descend into obsolescence with the same frequency—I will not here provide names of resources, except where needed to give a good example of the type of tool or resource that is needed.

Social network analysis (SNA) is used by digital humanities (DH) researchers to capture and quantify relationships between entities of any kind—normally between people in groups that have several "circles" of influence. Where network analysis has been performed around a number of things—like computers—social network analysis focuses on people. DH assignment 6 uses SNA to focus on relationships among historical people, as chosen by the individual students, themselves.

This DH assignment is loosely based on an excellent workshop session provided by Ruth Ahnert (Queen Mary University of London) and Sebastian Ahnert (University of Cambridge) at EMDA 2015.[1] In the general spirit of code reuse, I attribute to them the general approach to the tutorial (creating the edge file, the visualization, and the use of *networkx*[2] to do analysis in Python), and the tool choice to do these things in a classroom demonstration environment (since the tools they presented in the workshop are still, I believe, the best ones to use today). I have provided the edge file examples, the explanation of the process, and the code example, which generally follows their progression through the *networkx* methods but employs my own coding style. I thank them for providing this workshop to me, and see this chapter as a homage to their excellent work in the field of social network analysis.

With that understood, there are actually two layers of code reuse going on in this assignment: one layer having to do with the tutorial approach and code, and the second layer having to do with the Python library called *networkx*.

In this assignment, the students will be using an existing library in Python (*networkx*) to do social network analysis.

10.2 The Social Network Analysis Assignment

To do this assignment, you must either have the *networkx* library installed on your computer, or use a web programming tool (like Pythonanywhere.com,[3] which automatically has the *networkx* library installed) to write your code. Your instructor will determine which approach to take, and will announce it a few days before the tutorial session for this assignment.

There is some basic terminology which is used when doing social network analysis. The individual entities (in the case of this assignment, historical persons) that are being analyzed are referred to as *nodes*. The documented connection between two *nodes* is called an *edge*. Though the entire network of *nodes* and *edges* are referred to by a number of names (mostly differing in terms of how they are portrayed), in this assignment we will refer to it as a *graph*. In this assignment, you will create a social network *graph* by researching a historical figure of your choosing, typing the names of people associated with your historical figure (and others, by extension, as you will see below) into a csv (comma-separated values) file, and then analyzing that file both visually and by using algorithms provided in the *networkx* library.

[1]A special thanks to the Folger Institute of the Folger Shakespeare Library for supporting this research by hosting the "Early Modern Digital Agendas: Advanced Topics" institute in 2015 through a National Endowment for the Humanities grant. The Drs. Ahnert provided 1 day of activities involving Social Network Analysis at the institute, and were so impressive that their presentations were the focus of EMDA 2017, two years later.

[2]https://networkx.github.io/.

[3]https://www.pythonanywhere.com/.

Fig. 10.1 Small excerpt from my *edge* file created with Robert Hayden as the focus

	A	B
11	Robert Hayden	Pres. Jimmy Carter
12	Robert Hayden	Elinor Wylie
13	Robert Hayden	Countee Cullen
14	Robert Hayden	Paul Laurence Dunbar
15	Robert Hayden	Langston Hughes
16	Robert Hayden	Arna Bontemps
17	W. H. Auden	Christopher Isherwood
18	W. H. Auden	Chester Kallman
19	W. H. Auden	Igor Stravinsky

You will begin the assignment by choosing a person of historical significance in any area of your interest. Look up that person on Wikipedia, or any other information source. Scan the page and write down any names that are referred to as being in any way connected with your person—positively or negatively. For each of these names, whenever there is a link for that name in Wikipedia, follow that link and then do a similar scan on all connected persons' pages as well, writing down any names found therein. Go back and forth in this manner, and then also do the links of related persons in the linked pages as well. If some of the names found on these pages are the same, go ahead and record them anyway. Try to go at least three levels (or in Kevin Bacon terms, three degrees of separation) from the original person.

For example, if I was looking up Robert Hayden, the great American poet, and came upon the text, "… Hayden studied under W. H. Auden, who directed his attention to issues of poetic form, technique, and artistic discipline." [1], I would type Robert Hayden into column A of one row in the spreadsheet, and would type W. H. Auden into column B in the spreadsheet. I would also search for people with whom W. H. Auden were associated. A small excerpt of my *edge* file for Robert Hayden is provided in Fig. 10.1.

I would also follow links to look up Langston Hughes, Countee Cullen, etc., until I had more than 100 *edges* to process in my file, or 100 or more rows in the spreadsheet, that will eventually be saved as a csv file. Note, there will only be two columns in the spreadsheet, with a single name of the two people who have some kind of connection in each of the columns. There are two things to keep in mind as you make your *edge* file.

First, some names will likely appear more than once, associated with other names which might appear more than once. This kind of thing is normal. It also does not matter if a name is in column A or column B, or in both columns in different rows. The *networkx* library is very good about creating *nodes*, as long as the entities are spelled the same way. Which brings us to the next thing to keep in mind.

Secondly, you must make sure that the same entity (person) is represented in exactly the same way in any appearance of her or his name. For example, W.H.

Auden (with no space between W. and H.) and W. H. Auden (with a space) would appear as different *nodes*, due to the space inserted between said letters. It does not matter how you format a name, but it MUST be formatted and spelled exactly the same anywhere and everywhere you have that name in your edge file.

Once you have entered all of the connected names into your *edge* file, you are ready to do the remainder of the assignment. This first step is important, because all of the visualization and analysis that you do will use your *edges* as data. If you do a lousy job at creating your *edge* file, all of your analyses will be lousy as well. Remember: garbage in, garbage out.

The next step is to visualize your social network. You can use a number of visualization tools to do this, and there are tools being added constantly. You may have a favorite network visualization tool, or may be able to find a good one that is free to use on the web. If so, go to it! Most of the visualization tools on the web are easy to use. If you cannot find one, a tool that you might try using, if it is still available as you read this, is *Palladio*.[4] It has an easy-to-use interface and produces a good visualization of an *edge* file provided in the format that you produced above. Load your data into a network visualization tool and create a visualization of your social network. Figure 10.2 shows an example of a part of my network *graph* for the Robert Hayden network. I provided just this part to show the detail produced in the *graph*; you will want to make a graph of the entire network, which may zoom up to a scale where it is not possible to read all of the node names. If so, please also do screenshots of some zoomed-in details, like you see in Fig. 10.2. When you have tweaked your *edge* file and its accompanying visualization *graph* to your satisfaction, make a screenshot of this network and save it as a .png, .gif, or .jpg file.[5]

The final step is to modify the example code provided to analyze the network with at least the number of *nodes*, the degree of all of the entities (including the best connected from among all of the *nodes*), the shortest path between three sets of any two entities that you consider interesting, and the betweenness centrality value of three of the *nodes* of interest to you. All of these procedures and methods are described in the *networkx* documentation.[6]

When you've completed the assignment, you should compress these items into a zip file and submit the file on or before the assignment due date:

- Your edge file, as a csv file
- The screenshot of your visualized network, as a .png, .gif, or a .jpg file (or, if you cannot read the node names in the screenshot of the entire network, one or two additional image files zoomed to show representative detail)
- Your Python code that provides the required analysis mentioned above specifically written for your subject. In other words, please do not just submit the Python file of tutorial code provided. For example, if your subject is King John, you would

---

[4]At the time this book was printed, Palladio was available at http://hdlab.stanford.edu/palladio/.

[5]For instructions on how to do a screenshot on your computer, please consult the user's manual for your operating system, or do a simple web search.

[6]https://networkx.github.io/documentation/stable/.

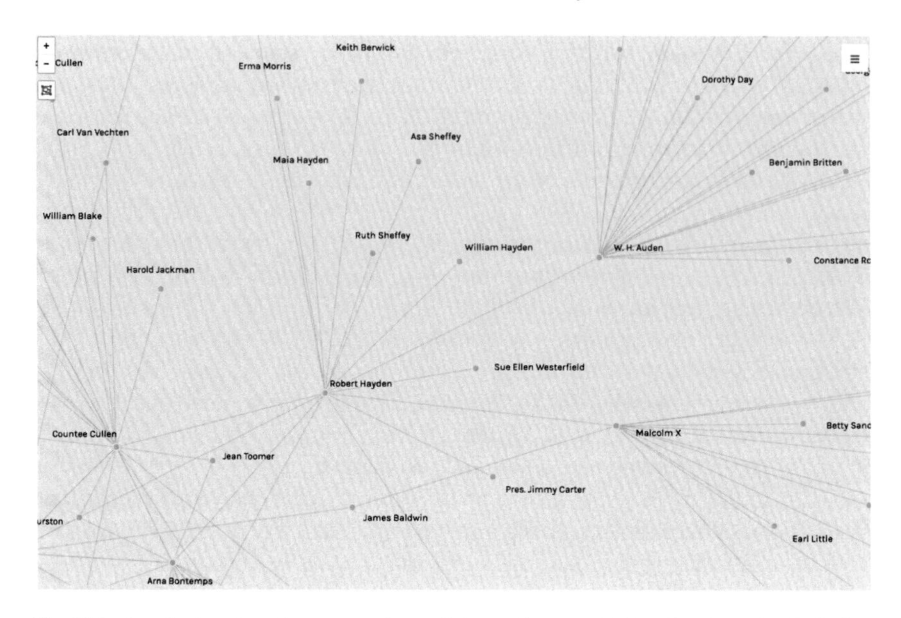

Fig. 10.2 Detailed section of my network *graph* focused on poet Robert Hayden using Palladio

not want to turn in code that attempts to find the shortest path between Robert Hayden and Langston Hughes.

10.3 Assignment Files and External Resources

edgeFileDemo.csv
edgeFileDemoWHeader.csv
Tutorial Code as an Atomic Exercise for this Assignment.

10.4 Skills Utilized in This Assignment

This assignment is designed to introduce students to importing and using library files as a way to employ tools. Digital humanities students are introduced to social network analysis and visualization using available tools.

New Programming Skills Utilized:

- Importing a specified library
- Instantiating objects of the specified library
- Running methods provided by objects in a specified library

- Exposure to Python dictionaries as a data structure
- From a demonstration of basic operations, modify the code to solve a problem
- Research the documentation of the specified library and add additional analysis of the student's choosing (for intermediate students and above).

New Digital Humanities Skills Utilized:

- Exposure to social network analysis and its place in historical research
- Familiarity with the issues involved with attempting to characterize and quantify relationships
- Ways to collect data to be used for SNA
- Different ways to analyze the relationships and draw inferences from them
- Discovering and using a good tool to visualize the relationships.

10.5 Assignment Management Techniques and Issues

This is the final assignment that I give my introductory programming students, near the last week of classes in the semester. At this time, they are usually completing papers and projects for other courses, and beginning to prepare for the final exam that I give in the course. Understanding that time is of a premium, I make this assignment an in-class tutorial, and design it so that, if they have made their *edge* files before the class session, they should be able to complete the entire assignment and submit their required materials by the end of the 75 minute class period. This allows them to finish all of their assignments during the last week of class sessions, and frees them up to prepare for the final exam.

One of the most difficult management problems associated with this assignment is the *networkx* library itself. The library is well-documented, including a good set of instructions for installing the library on different machine platforms. However, if I have a classroom of 26 novice students, most of whom bring in laptops with various and sundry versions of operating systems and levels of user access, I will have 26 installation problems that I need to run to ground. No way would an instructor be able to begin a class with installation of the library and then complete all of the other requirements within that single class session, unless students are working on homogenous systems, such as that provided by laptop programs. I have found that it is much easier to introduce students to a web IDE such as Python Anywhere, (see footnote 3) which at the time this book went to print, had a free service that satisfied our very limited needs in terms of consoles and files. The beauty of this service is that the *networkx* library is already installed—again, at the time of printing—students just needed to import the library and start using its methods. If something should occur that that website is no longer available, or no longer free, you might need to either lead the students through an installation of the *networkx* library in a prior class session, or find another web IDE that provides the library pre-installed. I hope I have provided you with a good "latchkey" solution in this chapter, but nothing is forever.

If you do use a service such as Python Anywhere, you would need to remind your students to get an account and make sure they can login at least one or two class sessions prior to the tutorial.

This assignment allows students to choose a historical person that interests them, so you will see a wide variety of subjects submitted by students for this assignment. I also occasionally have students approach me with other ideas, such as the characters of a book series (you can probably guess which one), or current pop culture figures and athletes. I usually don't have a problem letting them work in these areas, which are not historical per se, but do capture the passions of the students. It is certainly up to you how much you decide to stick to the historical aspect of this assignment.

In addition to making a decision regarding the *networkx* library installation, you will probably need to choose the visualization tool you will use in your tutorial. As mentioned in the assignment text, I have found the Palladio site to be easy to use and to provide a good network graph visualization, but you may have a favorite tool at your disposal. Or it may be that the Palladio site may go offline for a period of time in the future. If that's the case, from the recent trend in more free and open source software being distributed over a number of problems which only had proprietary software before, the odds are good that a good workable replacement tool will be available to you at that time.

I sell this tutorial class session late in the semester as a "no miss" session for students, telling them they will pretty much complete the assignment in class (assuming that they have brought a good *edge* file with them). For students who miss the class session, the description of the process, and perhaps providing the atomic tutorial code, will have to suffice. Many a good student who had to miss the class session due to some approved event or illness has struggled with this assignment, though, because there is a big difference between doing a tutorial along with an instructor, and trying to do the tasks while reading from a page on a learning management system. You will have to decide how much one-on-one tutoring you will want to do with these students (though that type of work could probably be handled pretty well by a teaching assistant).

Some students will have problems with file input—it may be from an encoding problem or a line ending problem. If the error cites an encoding problem, it could be that they need to add the final parameter encoding="utf-8" in their open statement for reading (more information about this problem is given in Chap. 7, and at various places on the web). It could also be that the students have not saved their *edge* files as csv files, but instead as xslx files. If the latter, these students will not be able to process the *edges*, since they are embedded in XML code that will not make sense in the context of their programs.

Other students might encounter a line ending problem: This usually arises when the student tries to input the file, and it says there is a parameter mismatch. You can determine that this is the problem if you view the input line with the student, and see that the entire file was input as one line. This often happens with Apple OS machines,

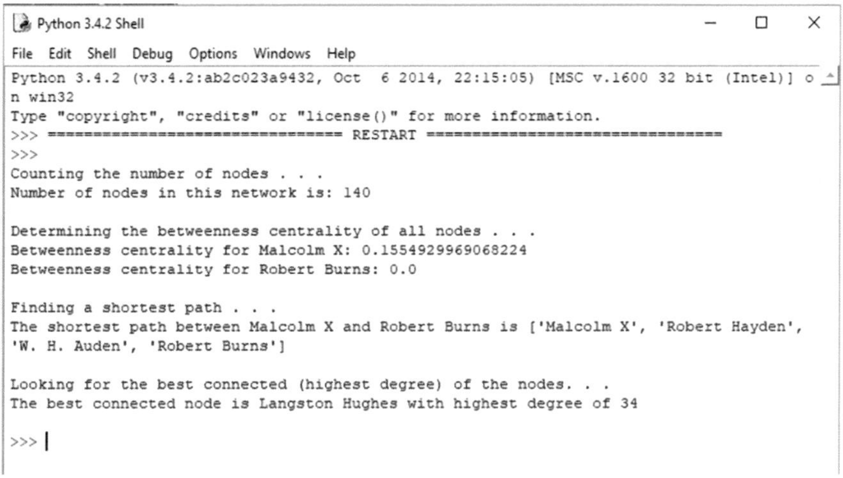

```
Python 3.4.2 Shell
File Edit Shell Debug Options Windows Help
Python 3.4.2 (v3.4.2:ab2c023a9432, Oct  6 2014, 22:15:05) [MSC v.1600 32 bit (Intel)] o
n win32
Type "copyright", "credits" or "license()" for more information.
>>> ================================ RESTART ================================
>>>
Counting the number of nodes . . .
Number of nodes in this network is: 140

Determining the betweenness centrality of all nodes . . .
Betweenness centrality for Malcolm X: 0.1554929969068224
Betweenness centrality for Robert Burns: 0.0

Finding a shortest path . . .
The shortest path between Malcolm X and Robert Burns is ['Malcolm X', 'Robert Hayden',
'W. H. Auden', 'Robert Burns']

Looking for the best connected (highest degree) of the nodes. . .
The best connected node is Langston Hughes with highest degree of 34

>>> |
```

Fig. 10.3 Screenshot of output from the analysis part of the SNA assignment

which handle line endings differently. For these students, they should be counselled to save their *edge* files as Windows CSV or MSDOS CSV files, and those problems often go away.

10.6 Atomic Code for This Assignment

Since this assignment is often done as an in-class tutorial, the starter code that I use in the tutorial session is recommended as atomic code for this assignment.

10.7 Expected Output from Student Work

An example of the type of edge file that might be submitted by students is shown in Fig. 10.1 above. An example of the screenshot of the network visualization may be seen by dropping the *edge* demo files into the visualization tool. A detailed (zoomed in) partial view of my visualization is provided in Fig. 10.2. An example terminal output expected from the students' adaptations of your tutorial code is shown below in Fig. 10.3.

10.8 Assignment Variations

10.8.1 For Novice Students (Taking an Introductory Programming Course)

If they have done the research to prepare their *edge* files, and bring them to the tutorial class session, novice students should be able to complete the tutorial, and change the code to adapt it to their specific subjects, within the class period. I usually ask them to upload their materials at the end of the class session, and then chase down the students who have not uploaded to make sure that they understand what is necessary to submit.

10.8.2 For Intermediate Students (Taking a Python Programming Course)

For intermediate students, in addition to doing all of the parts required of novice students, I also have them read the documentation in the *networkx* github site (cited elsewhere in this chapter) and choose an algorithm or analytical feature that interests them. Using the tutorials, examples, and other documentation, I have them employ this additional feature and print out the results. I also require them to provide a brief written summary of the algorithm or analytical feature that they chose, and what using that feature helped them learn about their networks.

10.8.3 For Advanced Students (Taking a Software Engineering or Capstone Course)

In addition to the tasks for the novice and intermediate students, I would also advise you to assign advanced students the additional task of creating their own visualization script to reveal the graph to users visually. This will provide a good refresher on graphics processing, and will provide them with additional strategies to help them reveal relationships between entities.

10.8.4 For Secondary School Students

Since the novice version of this assignment is built on a guided tutorial approach, secondary school students should be able to follow along and do this exercise as well. It may not be necessary to have them perform additional operations and analysis on their own, based on your confidence in their abilities. Perhaps just doing the tutorial along with you will suffice to meet your student learning objectives.

10.8.5 For Digital Humanities Students

Digital humanities students will likely be enthralled with this tool, and will need little prodding to do their own exploration in their areas of interest. In this case, more time could be spent allowing these students to determine their subjects and gather data, with perhaps giving them an opportunity to do some follow-up reflection of insights gained about the subject matter through the experience of doing it. DH students should also be encouraged to take the time to install the *networkx* library on their laptops and home computers, as this will give them optimal access to the library for further exploration beyond the end of the course.

Reference

1. Hayden R.: Wikipedia. https://en.wikipedia.org/wiki/Robert_Hayden (2018). Accessed 20 June 2018

Chapter 11
Conclusion

Abstract This chapter summarizes the six Digital Humanities assignments provided in the book, including some suggestions for two additional analytical methods (spatial analysis and topic modeling) which were not implemented as assignments, but could be. Some suggestions for how these two additional methods might be implemented successfully in the classroom are provided. The chapter ends with the assertion that learning to work with humanities data may make a positive difference in the lives of all students in a course that uses digital humanities assignments in an introductory programming course.

Keywords Digital-humanities programming · Introductory programming
Spatial analysis · Topic modeling · Humanities data · Mindful computing

Hopefully this book has provided you with effective digital humanities (DH) assignments that will help your students learn fundamental Python programming concepts while also familiarizing them with some of the analytical methods utilized by DH programmers. The assignments presented in this book have been designed for successful implementation in classrooms with a number of different makeups, from novice to advanced students, with some special nuances for secondary school and DH (non-computer science) students.

There are two additional analytical methods that were discussed in Chap. 3 (Digital Humanities—Special Considerations for the Programmer) but were not adapted into assignments, because the level of programming skill required to do even the most basic operations in these methods is beyond that of the average introductory programming student (even near the end of that novice's full semester of study). These additional methods, discussed below, may be successfully implemented as assignments for intermediate and advanced programming students. Though I have not actually embodied them into assignments (since my only programming courses are taught at an introductory level), here are some thoughts about how you might go about it if you are ambitious and have some time on your hands:

Spatial Analysis (includes Geospatial Analysis)—a systematic method of analyzing an artifact in reference to its location, both in relation to other physical characteristics of the Earth and in relation to other artifacts. This is another area,

B. Kokensparger, *Guide to Programming for the Digital Humanities*, SpringerBriefs in Computer Science, https://doi.org/10.1007/978-3-319-99115-3_11

like Social Network Analysis, where stable libraries have been created for Python users who want to programmatically explore and implement mapping and spatial analysis activities. The currently-available libraries each have their own strengths and weaknesses, and having little to no direct experience with them, I would be foolish to make an opinion regarding which one may be best for you or your students. Suffice it to say that any programming dealing with spatial analysis is beyond the scope of this book, but if you work to implement a spatial analysis assignment in your class, you will have plenty of experience with the various Python libraries available for this task. A good approach for this kind of assignment may be a tutorial approach, with requirements tied into students' individually-developed data sets, much like the Social Network Analysis assignment in Chap. 10.

Topic Modeling—a systematic method of discovering summarizing topics in given artifacts and determining how they relate to other topics in the same or other artifacts. I have used topic modeling software both as an end in itself (reporting upon the output), and as a feedback mechanism (feeding the output back into the input to the same topic modeling tool, or as input to other DH tools). The kind of textual manipulation that occurs when cleaning and aggregating csv (comma separated values) files that are output from DH tools, and/or providing input for other DH tools, might be a useful assignment for intermediate and advanced students, or perhaps at the end of the semester for a talented group of novice students. Since the requirements for such an assignment would vary greatly depending upon the field of application (historical, literary, archeological, etc.), and the tools used for input and output would vary as well, I cannot give much more help but to point you to your own academic interests as an instructor. If you find a certain approach to an academic question interesting, and are able to involve your students in the collection, cleaning, and analysis of the relevant data, your students cannot help but learn a great deal from you, and from the data, in the process.

As the DH field is ever-changing, with constantly shifting topics and interests, I cannot help but think that a book such as this one, published ten years later, may have different assignments focusing on areas of interest to humanities researchers that are currently unpredictable by anyone in the field. As a teaching professional in your field, you may come up with DH assignments of your own that work well for you in the classroom. If so, I hope you share them with others, and with me.

I hope you have learned, both in reading this book and implementing its assignments, that when you teach students how to program by engaging the data in the Human, the Human in the data is bound to emerge and teach them something about themselves, and their world. Where these learning opportunities may be seen by computer science students as disruptions, or even failures, it is my hope that, in learning how to program for the humanities, they instead are seen by your students as opportunities to engage more closely with the world as it is, not as they think it should be. In doing so, we are not making humanities more like mindless computation, we are making computation more mindful, and perhaps even soulful.

I hope that you and your students have also learned that DH programming has its own rewards, beyond the great opportunities that lie in careers that work with human artifacts and the connections among them, as extensions of the people and places where they were created and valued. Few computer programmers get the opportunity to work with things that have made a difference over time in how we, as humans, view and alter our world. Beyond training students to program for the humanities, we are training them to contribute to this ongoing expression of the Human.

Made in the USA
Middletown, DE
12 March 2019